Wilhelm Ostwald, George McGowan

The Scientific Foundations of Analytical Chemistry Treated in an Elementary Manner

Wilhelm Ostwald, George McGowan

The Scientific Foundations of Analytical Chemistry Treated in an Elementary Manner

ISBN/EAN: 9783337058975

Printed in Europe, USA, Canada, Australia, Japan

Cover: Foto ©berggeist007 / pixelio.de

More available books at **www.hansebooks.com**

THE SCIENTIFIC FOUNDATIONS

OF

ANALYTICAL CHEMISTRY

THE
SCIENTIFIC FOUNDATIONS
OF
ANALYTICAL CHEMISTRY

TREATED IN AN ELEMENTARY MANNER

BY

WILHELM OSTWALD, Ph.D.

PROFESSOR OF CHEMISTRY IN THE UNIVERSITY OF LEIPZIG

TRANSLATED WITH THE AUTHOR'S SANCTION

BY

GEORGE M'GOWAN, Ph.D.

London

MACMILLAN AND CO.

AND NEW YORK

1895

All rights reserved

DEDICATED

TO

Johannes Wislicenus

IN

AFFECTIONATE RESPECT AND FRIENDSHIP

AUTHOR'S PREFACE

ANALYTICAL Chemistry, or the art of recognising different substances and determining their constituents, takes a prominent position among the applications of the science, since the questions which it enables us to answer arise wherever chemical processes are employed for scientific or technical purposes. Its supreme importance has caused it to be assiduously cultivated from a very early period in the history of chemistry, and its records comprise a large part of the quantitative work which is spread over the whole domain of the science. There is, however, a remarkable contrast between the extent to which the *technique* of analytical chemistry has been elaborated and its scientific treatment. Even in the best works on the subject the latter is almost entirely confined to the giving of equation-formulæ, which show the results of the chemical reactions in question *in the ideal limit cases*. That as a matter of fact, instead of the supposed complete reactions, incomplete ones leading to a state of chemical equilibrium take place, that there is no such thing as a perfectly insoluble substance, and that absolutely exact methods of separation and estimation are an impossibility—remains not merely unknown to the student, but also occurs less frequently, I fear, to the mind of the accomplished analyst than is to be desired in the interests of a sound criticism of analytical methods and their results.

Analytical chemistry thus fills the subordinate but at the same time indispensable position of handmaid to the other branches of our science. While we everywhere find the liveliest activity with regard to the theoretical arrangement of scientific material, and observe that questions of this kind always arouse far more interest than purely experimental problems, analytical chemistry is content with fashions of theory which have long been discarded elsewhere, and sees no harm in presenting its results in a shape which has really been antiquated for the last half century. Thus we find it considered permissible to give at the present day (for example) K_2O and SO_3 as the constituents of potassium sulphate, in accordance with the electro-chemical dualism of 1820; and the case is made no better by the fact that chlorine is brought into the report of an analysis as chlorine, and its 'oxygen equivalent' therefore deducted from the sum total.

We may, however, take it for certain that when such a striking and pronounced custom holds its own for so long, there must be good grounds for it. And it must be added, without any circumlocution, that a scientific foundation and system of analytical chemistry have hitherto failed us *because the general knowledge and laws necessary for these have not been at the disposal of scientific chemistry itself.* It is only within the last few years, *i.e.* since the development of the general theory of chemical reactions and states of equilibrium, that it has become possible to elaborate a theory of analytical reactions. It will be my endeavour to show in the following pages to what a great extent light has been thrown from this quarter upon long familiar and daily recurring chemical phenomena.

LEIPZIG, *May* 1894.

TRANSLATOR'S PREFACE

It is only necessary for me to add a word or two with regard to this translation. Professor Ostwald has made here and there an emendation upon the original German edition, which appeared about a year ago, and has added a section of a chapter upon electro-analysis. At my request he has further been good enough to go carefully through the revised proof-sheets. I should also like to take this opportunity of expressing my hearty thanks to Professor William Ramsay for much kind help and criticism.

GEORGE M'GOWAN.

July 1895.

CONTENTS

PART I—THEORY

CHAPTER I

The Recognition of Different Substances

	PAGE
General Considerations with regard to the Foundations of Chemical Analysis	1
Properties of Substances	2
Reactions	4
The Graduation of Properties	5
Colour and Light	7

CHAPTER II

The Separation of Substances

General Considerations	9
Separation of Solids from Solids	10
Separation of Liquids from Solids: Filtration	13
Washing of Precipitates	15
Theory of the Washing of Precipitates	17
Adsorption Phenomena	18
The Enlargement of the Crystalline Grains	22
Colloidal Precipitates	24
Decantation	27

	PAGE
SEPARATION OF LIQUIDS FROM LIQUIDS	28
SEPARATION OF GASES FROM SOLIDS OR LIQUIDS	28
SEPARATION OF GASES FROM ONE ANOTHER	29

CHAPTER III

PHYSICAL METHODS OF SEPARATION

GENERAL CONSIDERATIONS	31
THE THEORY OF DISTILLATION	32
DISTILLATION OF MIXED LIQUIDS	34
SEPARATION BY SOLUTION	38
SOLUTIONS OF GASES	39
THE DRYING OF GASES	41
TWO NON-MISCIBLE LIQUIDS: THE THEORY OF THE EXTRACTION OF A DISSOLVED SUBSTANCE FROM ONE SOLVENT BY SHAKING THIS UP WITH ANOTHER	41
SOLUTIONS OF SOLIDS	43
THE SEPARATION OF SEVERAL SOLUBLE SUBSTANCES	45

CHAPTER IV

CHEMICAL SEPARATION

§ 1. THE THEORY OF SOLUTION	46
INTRODUCTION	46
THE STATE OF SUBSTANCES IN SOLUTION	47
IONS	48
THE VARIETIES OF IONS	49
SOME FURTHER DETAILS	54
§ 2. CHEMICAL EQUILIBRIUM	56
THE LAW OF MASS-ACTION	56
APPLICATIONS	59
COMPLEX DISSOCIATION	60
GRADUAL DISSOCIATION	61

CONTENTS

	PAGE
SEVERAL ELECTROLYTES TOGETHER	62
ACIDS AND THEIR OWN SALTS	64
HYDROLYSIS	66
HETEROGENEOUS EQUILIBRIUM: LAW OF DISTRIBUTION	68
§ 3. THE COURSE OF CHEMICAL REACTIONS	70
THE VELOCITY OF REACTION	70
INFLUENCE OF TEMPERATURE	71
CATALYSIS	71
HETEROGENEOUS STRUCTURES	72
§ 4. PRECIPITATION	73
GENERAL CONSIDERATIONS	73
SUPER-SATURATION	73
THE SOLUBILITY-PRODUCT	75
SOME PRECIPITATION-REACTIONS	78
THE REDISSOLVING OF PRECIPITATES	81
§ 5. REACTIONS ATTENDED WITH THE LIBERATION OR ABSORPTION OF GAS	85
THE LIBERATION OF GAS	85
THE ABSORPTION OF GAS	88
§ 6. REACTIONS ACCOMPANYING THE EXTRACTION OF A DISSOLVED SUBSTANCE FROM AQUEOUS SOLUTION	89
INFLUENCE OF THE IONIC STATE	89
§ 7. THE ELECTROLYTIC METHOD	90

CHAPTER V

THE QUANTITATIVE DETERMINATION OF SUBSTANCES

GENERAL CONSIDERATIONS	99
PURE SUBSTANCES	101

	PAGE
BINARY MIXTURES	104
INDIRECT ANALYSIS	106
TERTIARY MIXTURES, ETC.	108
METHODS OF TITRATION	109

PART II—APPLICATIONS

INTRODUCTION	115

CHAPTER VI

THE HYDROGEN AND HYDROXYL IONS

ACIDS AND BASES	117
THE THEORY OF INDICATORS	118
THE PRESENCE OF CARBONIC ACID	121
POLYBASIC ACIDS	122

CHAPTER VII

THE METALS OF THE ALKALIES

GENERAL PROPERTIES	124
POTASSIUM, RUBIDIUM AND CÆSIUM	124
SODIUM	126
LITHIUM	127
AMMONIA	128

CHAPTER VIII

THE METALS OF THE ALKALINE EARTHS

GENERAL PROPERTIES	130
CALCIUM	131

	PAGE
STRONTIUM	133
BARIUM	135
MAGNESIUM	136
APPENDIX	138

CHAPTER IX

THE METALS OF THE IRON GROUP

	PAGE
GENERAL PROPERTIES	139
IRON	140
CHROMIUM	144
MANGANESE	146
COBALT AND NICKEL	148
ZINC	150

CHAPTER X

THE METALS OF THE COPPER GROUP

	PAGE
GENERAL PROPERTIES	153
CADMIUM	155
COPPER	156
SILVER	159
MERCURY	161
LEAD	165
BISMUTH	167

CHAPTER XI

THE METALS OF THE TIN GROUP

	PAGE
GENERAL PROPERTIES	168
TIN	169

xviii FOUNDATIONS OF ANALYTICAL CHEMISTRY

	PAGE
ANTIMONY	171
ARSENIC	174

CHAPTER XII

THE NON-METALS

GENERAL PROPERTIES	176
THE HALOGENS	177
CYANOGEN AND THIOCYANOGEN	182
THE MONOBASIC OXYGEN ACIDS	184
THE ACIDS OF SULPHUR	187
CARBONIC ACID	192
PHOSPHORIC ACID	194
PHOSPHOROUS AND HYPOPHOSPHOROUS ACIDS	198
BORACIC ACID	200
SILICIC ACID	201

CHAPTER XIII

THE CALCULATION OF ANALYSES	204

PART I
THEORY

CHAPTER I

THE RECOGNITION OF DIFFERENT SUBSTANCES

1. *General Considerations with regard to the Foundations of Chemical Analysis*

THE first step in the solution of the problem—how to determine the nature of any given kind of matter—follows from the knowledge of its properties, as these appeal to our senses. We find no difficulty, for instance, in pronouncing one particular substance to be sulphur; if it has a yellow colour and a low specific gravity, and if it burns with a blue flame and an odour of sulphur dioxide, leaving at the same time no residue, we feel assured that it can be nothing else.

In coming to a conclusion of this kind we make use of various empirical data, which seem to us for the most part self-evident, and which do not therefore receive outward expression. Thus, the number of properties appertaining to one particular substance is indefinitely great; it is therefore not possible to state at once definitely, with regard to two objects, that they agree in *all* their properties throughout. But, as the result of an unexpressed induction of very wide applicability, we hold any such proof to be superfluous, for we know *that if two substances agree perfectly in some of their properties, they will agree in all.*

This empirical statement is nothing else than an expression of the corresponding fact that the number of different substances is limited and finite. Since the differences in the varieties of matter consist merely in differences in their properties, these properties and their values can obviously not be combinable in an unlimited degree, otherwise we should have an infinity of substances.

This allows to analytical chemistry a most desirable freedom in the selection of those properties which it utilises for the characterisation of different substances; almost any one is as good as another for the purpose, so far as principle is concerned, but the choice actually depends upon the ease and certainty with which the properties in question can be observed and measured. In many cases the determination of one property alone is sufficient for the recognition of the substance; as a rule, however, several such determinations are combined, in order to avoid any possible confusion which might arise from the (merely) approximate accuracy of the methods of determination and measurement employed. The probability of error diminishes very rapidly with the increase in the number of independent tests.

2. *Properties*

If, under the term *properties* of any given object, we mean all the relations in which it can be made to appeal to our senses and to our measuring apparatus, we may at present exclude all those properties which can be brought to light and altered at will, such as outward form, position and motion, illumination, communicable electric condition, temperature, and so on. Further, such properties as are uninfluenced by material or chemical changes will not serve for the recognition of definite substances. This applies more especially to the *mass* of bodies and to their proportional *weight*. Thus only those properties are applicable here which change with the nature of the substances themselves, but which cannot be arbitrarily altered in any one substance.

Every property can be defined numerically, and can show an infinite diversity of particular cases between the limits of its values. As a matter of fact, however, this infinity reduces itself practically to a finite number of distinguishable cases, since the means of determining the numerical values are always of approximate accuracy only. Progress in the art of measurement thus means a continuous enlargement in the number of distinguishable steps, without the theoretical infinity ever becoming attainable. At the same time refinement in measurement has now been carried to such a degree in the case of many properties that the number of distinguishable cases far exceeds that which actually occurs.

The properties which are made use of for analytical purposes may be divided into the two following groups : *properties of condition or state*, and *properties of reaction*. The former are an invariable attribute of the object in question, and are capable of direct observation and measurement at any time. Among them we have, for instance, state of aggregation, colour, specific gravity, and so on. Other properties first come to light when the object is brought under special conditions which are different from the ordinary ones. In this way changes of state are brought about, or reactions are induced, which are characteristic of the particular kinds of matter. It follows from the nature of things that the second group of properties is by far the larger and more diversified; the reaction-properties thus play a far more important part in analytical chemistry than the properties of condition.

3. *Reactions*

Reactions are called forth by changes induced in the conditions under which the object is placed, and such changes may be divided into physical and chemical. The most important physical change, so far as our present purpose is concerned, is that of temperature, and the behaviour of substances when heated has always furnished one of the most valuable aids to chemical analysis. Other physical changes, such as those of pressure and electrical condition, come into question much less often. But far more varied are the chemical changes which we are able to bring about in the conditions of existence of any given substance. This is generally effected by bringing it into contact with other substances. The contact is most complete between two gases or two miscible liquids, less perfect between two substances of different states of aggregation, and least perfect between two solids. It follows from this that the liquid state is by far the most convenient for the purpose just named, all the more since comparatively few substances can be converted into gas. Thus the aim of the analytical chemist, so far as regards bringing about chemical changes, is directed in the first instance to the production of the liquid state, either by fusion or solution.

In reality the recognition of substances by reactions leads us back to the recognition of them through properties of condition; only these no longer apply to the original object, but to that into which it has been transformed by the reaction. If we notice, for instance, that a precipitate is formed upon the addition of one liquid to another, the observation rests upon the fact that—under the altered conditions—a substance in the solid state of aggregation results. And the same may be said for all reactions, so that the investigation of the nature of the properties of condition is of importance for both groups.

4. *The Graduation of Properties*

It has been already stated that, so far as principle goes, any one property of condition may be made use of for the recognition of substances. The distinguishing of different kinds of matter is invariably based upon quantitative differences of the property in question. The task of determining such differences is, however, one of very varying difficulty, according to the nature of these; and it is usually some properties, whose differences can be easily and quickly established, which are taken into consideration. Among those the state of aggregation deserves first place and colour the second. Whether a substance is solid, liquid, or gaseous, and what colour it has, can usually be seen at a glance; these properties are therefore to be considered first in discussing the question as to how different substances are to be recognised.

As is well known, we have transition stages

between the three states of aggregation, but these concern us little at present. The gradual change from the gaseous to the liquid state is the result of a pressure which is higher than the critical pressure. Since, however, the critical pressures of substances vary—roughly speaking—between twenty-five and a hundred atmospheres, these transitions do not come into play under the conditions of ordinary analytical operations. The changes between the solid and the liquid states are of more importance. These are either sudden, as in the melting of ice, or gradual, as in the fusion of glass. The latter occurs when the solid body is amorphous, while the former is peculiar to crystalline substances.

The above transition states may be still further subdivided by the eye alone into several grades, by making use of very simple methods. Thus we can distinguish between mobile, fluid, viscous, sticky and solid substances, although it is not possible to extend the characterisation beyond four or five stages without other aids.

In the case of solids it is often quite easy to decide whether they are amorphous or crystalline, especially if we are dealing with fairly large fragments; amorphous bodies show a conchoidal fracture and their surfaces are uneven, while those which are crystalline give a fracture of larger and smaller plane surfaces. The point cannot always be decided with certainty by the naked eye when dealing with powders; these require either the pocket lens or the microscope.

5. *Colour and Light*

The colour of substances is a characteristic of very wide applicability. The fact that relatively small differences in the wave-lengths of reflected light affect the eye as differences in colour has for its result that this quantitative difference is converted into a series of qualitatively distinct, though at the same time continuous, steps; we are thus enabled to distinguish with ease ten, twenty, or even more grades of colour, and to make use of these for purposes of recognition. We have, however, to bear in mind here that the surfaces of coloured bodies radiate as a general rule a mixture of two different kinds of light—that which is coloured through absorption, and which proceeds more or less from the interior, and that which is thrown back by surface reflection, the latter being usually white light. The relation between these depends upon a number of conditions, more especially upon the degree of subdivision and upon the difference between the refraction-coefficient of the substance and of the medium surrounding it. According to the amount of white surface light, the colour of a substance may vary between white and a very dark tint, which often approximates to black; it is therefore necessary as a rule, when talking of the colour of anything, to specify the conditions under which that colour is to be observed (*e.g.* whether the substance is compact or powdery, or floating in a liquid). Most of the cases in point, which occur in analytical chemistry, apply to powders deposited from water, *i.e.* obtained as precipitates in chemical reactions.

Besides the colours of substances which are produced by illumination in white daylight there is another colour phenomenon of importance for analytical chemistry,—coloured flames. These result when certain substances are heated in as nearly non-luminous a flame as possible (such as the flame of a *Bunsen* burner or a spirit lamp), whereby the latter volatilise and emit light consisting of a limited number of rays of definite kinds and therefore of definite colours. This phenomenon can be observed in its simplest form, *i.e.* so far as colour goes, by the naked eye; it attains, however, to an infinitely higher degree of diversity if the light of such a flame is separated into its constituent parts by the spectroscope, when it becomes one of the most thorough and certain aids towards the recognition of those substances which yield coloured flames.

In addition to the readily apparent properties that have just been detailed, there are many others which may be made to assist in the recognition of different substances, but they are all much slower and more difficult of application, and need not therefore be considered practically here.

CHAPTER II

THE SEPARATION OF SUBSTANCES

1. *General Considerations*

FROM what has been said in the foregoing chapter it is apparent that the task of recognising any given substance, *i.e.* of identifying it with one already known, is always more or less easy of accomplishment, and merely presupposes a practical system in the choice and tabulation of the properties which are made use of for this recognition, so as to reduce the labour to a minimum and to attain to the highest possible accuracy. But the problem becomes far more complicated when we have to deal, not with a simple substance, but with a mixture; *separation* must here precede *recognition*, and the first-named operation is naturally much the more difficult of the two.

In order to be able to separate one substance from several others, it is necessary to have the first in a condition in which it is detached from the others by a surface of separation or unstable equilibrium. Such surfaces of separation occur in the first instance and chiefly when there are different states of aggregation, although they are not necessarily excluded when the state of aggregation is the same. The plan of methods of separation is best referred to the different states of aggregation, and we have thus to consider the following cases:—

(*a*) Solids from solids.
(*b*) Solids from liquids.
(*c*) Liquids from liquids.
(*d*) Solids or liquids from gases.
(*e*) Gases from gases.

The separation of substances is always a mechanical operation; a so-called chemical separation consists in transforming the substances in question, by chemical means, into others which can be separated mechanically.

2. *Separation of Solids from Solids*

The principle upon which these separations are based consists in allowing forces to act upon one or other of the constituents, which transport the latter to a spot from whence it can be removed.

Differences in the specific gravity of substances constitute the property which is the most widely applicable for this purpose. If a mixture of two solids be rubbed up in a liquid whose specific gravity lies between their own, the lighter one will rise to the top and the heavier one sink to the bottom, and so a separation will be effected. Should the specific gravities of the solids be known beforehand, that of the liquid can be regulated accordingly. But should this not be the case, then we have to begin with a liquid which is denser than both, and lower its specific gravity, *e.g.* by the addition of a lighter liquid, until the wished-for separation is brought about.

If more than two solids are present, the same

method may be followed of gradually lowering the density of the liquid. Proceeding in this way by short steps, the heaviest solid falls to the bottom first and can be removed; then the next heaviest, and so on.

Aqueous solutions of potassium-mercury iodide, barium-mercury iodide, cadmium borotungstate, and similar salts are suitable liquids for substances insoluble in water. For those that are soluble methylene iodide may be used, and it can be diluted with light liquids such as xylene. The specific gravity of such liquids does not, however, much exceed 3, so that solids of greater density cannot be separated by their means. In some cases molten substances of higher specific gravity may be employed for these latter.

A similar procedure, though a much less perfect one, consists in elutriation or "washing." This method of separation is based upon the circumstance that solids in a more or less fine state of division sink the more quickly to the bottom of a liquid the denser they are. A stream of liquid thus carries away a preponderance of the less dense constituents. The rapidity with which suspended particles sink depends, however, not merely upon their specific gravity, but also in a very great degree upon their size, the smaller ones subsiding more slowly than the larger. As a result of this complex relation the process is unsuited to exact separations. In order to derive the greatest practicable benefit from it, the particles to be washed should be made as nearly alike in size as possible, which is best achieved by grinding the whole to a fine powder. The method can only be followed on a practical scale when the differences in specific gravity are somewhat marked.

Beyond hydrostatic forces [1] there are none known which are of general applicability for the separation of solids. In particular cases, however, other forces—more especially magnetic—are made use of for such separations. Thus, particles of iron can be removed from a mixture by means of a magnet (for substances which are but weakly magnetic a powerful electromagnet is employed). Magnetic and hydrostatic forces can also be conjoined.

Electrostatic forces can likewise be utilised here. Mixtures of different substances in powder are electrified on shaking, the one constituent becoming positive and the other negative. If such a mixture is now brought into contact with an electrified nonconductor, *e.g.* a rubbed ebony plate, the oppositely charged particles are attracted to it and the others are repelled. I am not aware whether any application has been made of this for purposes of separation.

Still another mode of separating solids might be based on the fact that in a non-homogeneous electric field the substances with the higher dialectric constants are driven to those spots where the intensity of the field is greatest. No application has been made of this.

The process of separating a mixture of two solids by treatment with a solvent, in which one of the constituents is soluble and the other insoluble, does not come in here. It rests upon the establishment of two different states of aggregation for the substances under treatment, and therefore belongs in principle to the next section.

---

[1] Hydrostatic forces can be rendered much more efficacious by being joined to centrifugal force, which likewise finds frequent application. This, however, introduces no new principle into the procedure.

3. *Separation of Liquids from Solids—Filtration*

The process of separating a liquid from a solid is termed *filtration*, and depends upon the use of a porous medium whose pores are smaller than the particles of the solid. Since the mixture is allowed to exert pressure upon the walls of the filter, the liquid is driven through while the solid is retained.

Of all the methods of separation used in analysis, filtration is the one which is most applied, since it is the easiest to carry out and to control. The separation of gases from liquids and solids is indeed simpler in theory, and requires almost no apparatus; but the necessity of using large closed vessels when dealing with gases makes this much more troublesome than the handling of liquids and solids. For this reason separations in practical analysis are always reduced, if possible, to the two latter. Many different materials can serve for the porous medium, but only paper and asbestos need concern us here. The larger the particles of the solid are, the larger may the pores be; and, since filtration proceeds with greater rapidity when the particles of the precipitate are large, the aim of the worker is always to get his precipitate into this state, so far, that is, as other circumstances will permit. A very effective mode of increasing the size of the particles of a fine precipitate is to allow them to remain in contact with the liquid in which they are formed for a considerable time. The result of this is that a recrystallisation is brought about, by which the finer particles disappear and the coarser are enlarged at their cost, and this proceeds more quickly the higher the temperature. Under like conditions the fine granules of amorphous precipitates coalesce to larger ones. Hence the usual practice of digesting a precipitate in its liquid before proceeding with the filtration.

The rate of filtration varies with the size of the pores, the pressure and the temperature, and it increases simultaneously with all three factors. The size of the pores depends not merely on the original condition of the porous medium, but also in a great degree upon that of the powder. Very fine precipitates narrow the pores of the filter to a marked extent and thus retard the filtration, this constituting a further reason for the production of as large particles in precipitates as possible.

The force of gravity usually supplies the needful pressure. The latter can, however, be increased either by raising the level of the unfiltered liquid above that of the filter or by lowering the level of the filtrate beneath it. The first of these procedures is the simpler to carry out on a technical scale, but it is not as a rule very applicable in analysis, since the amount of liquid is usually too small for the purpose, especially towards the end of the filtration. The second method necessitates the shutting out of all air from the rim of the filter to the bottom of the attached tube, and thus requires some care; it is usually carried out by joining a long glass tube to the funnel.

Since the hydrostatic pressure depends upon the height of the column of liquid alone and not on its width, it is advisable to have as narrow an extension tube as possible. The friction of the liquid furnishes a limit in this respect, being inversely proportional to the fourth power of the diameter of the tube; it does not do therefore to have a tube less than a few millimeters in diameter. On the other hand, it is quite useless to have one wider than this.

The filtration pressure may be still further augmented by utilising the pressure of the air. And, here again, there are two ways of carrying this out, viz. either by increasing the pressure upon the filter or by diminishing it underneath. Since it is far more important for our purpose that the filter rather than the filtrate should be easily got at, the second procedure is almost invariably adopted. Filtration under diminished pressure was elaborated by Bunsen more especially, down to nearly its last detail, and is applied daily in every laboratory.

Lastly, filtration pressure may be increased at will by mechanical means, through the use of pumps, presses, etc. Appliances of this nature are of great value in technical working, when large quantities of liquids requiring filtration have to be handled, but they are very seldom used in analysis.

The third factor in accelerating filtration is temperature. Since the motion of a liquid in the pores of a filter depends upon its internal friction, the very great influence of temperature upon this property makes itself felt here. Thus, for example, the internal friction of water at 100° is less than one-sixth of what it is at 0°. Hence the rule, always to filter as hot as other circumstances will permit.

4. *Washing of Precipitates*

The separation of the liquid from the solid is not, however, complete at the end of the filtration proper, seeing that a part of the former still remains behind moistening the latter; the amount of liquid thus entangled is approximately proportional to the surface of the moistened precipitate, and therefore augments very rapidly with increasing fineness of the latter. In addition to this there is the liquid which is retained in the spaces between the particles of the powder by capillary attraction. To complete the separation, therefore, we require to wash the precipitate after filtering it off, and thus to displace the liquid in question by some other suitable one (usually water). Several things have to be taken into account in considering the theory of washing precipitates, the most important of these being the adsorption phenomena, *i.e.* the adhesion of substances in solution to solid surfaces. Further, many precipitates tend to "pass through the filter" in the course of washing. This arises from the property of colloidal substances to become broken up in pure water, while in solutions of salts they remain coagulated, and therefore in a fit condition for filtering. We have thus the empirical rule for this case—to wash with some suitable salt solution instead of with pure water. The theoretical discussion of all these phenomena will follow later.

The residue of the wash liquid which moistens the precipitate and which is retained by capillary attraction is finally got rid of by drying. And we have to remember here that the vapour pressure of the exceedingly thin films of the moistening liquid is much less than that of the same liquid in the free state. The drying temperature must therefore be raised far beyond the boiling-point of the liquid in order to practically get rid of the last traces of the latter, and the finer the powder the higher must the temperature be. Colloidal substances require the highest temperatures.

5. *Theory of the Washing of Precipitates*

If a represent the quantity of the liquid under filtration which remains behind with the precipitate, and also the residual quantity of the liquid used for washing (assuming that the latter is mixed uniformly with the precipitate each time), then, after m liquid has been poured on, the total amount of liquid will be $m + a$, and the original quantity will have been diluted to the $(m+a)$th degree. Again, if x_0 represents the concentration of the substance in the original solution which has to be displaced, its absolute amount before washing has commenced is ax_0. After the addition of a quantity m of the washing liquid, this concentration is reduced to the fraction $x_1 = \dfrac{a}{m+a} x_0$, and when this liquid in its turn has drained through the filter until only the quantity a remains mixed with the precipitate, the absolute amount has gone down to $ax_1 = \dfrac{a}{m+a} \cdot ax_0$. A second addition of washing fluid gives the concentration $x_2 = \dfrac{a}{m+a} x_1 = \left(\dfrac{a}{m+a}\right)^2 x_0$ and the absolute residual amount $ax_2 = \left(\dfrac{a}{m+a}\right)^2 ax_0$, until, after n washings, the residue of original liquid remaining with the precipitate is

$$ax_n = \left(\dfrac{a}{m+a}\right)^n ax_0.$$

It follows from this formula that, for an equal number n of washings, the residue ax_0 will be smaller the smaller the fraction $\dfrac{a}{m+a}$, that is, the more perfectly the precipitate is allowed to drain (whereby a is diminished) and the more washing liquid m is used each time. Should the latter, for example, amount to nine times as much as the original moistening solution, and should 1 gramme of foreign substance be mixed with the precipitate to begin with, then after four washings only $\left(\frac{1}{10}\right)^4 g$, *i.e.* 0·0001 grm., of the impurity would remain.

The answer to the question—how can one best wash a precipitate with *a given amount of liquid*—is somewhat different. To work this out requires the use of the differential calculus, and it will therefore be omitted here; the result, however, shows that it is better to wash many times with small quantities of liquid than a few times with large quantities.

6. *Adsorption Phenomena*

But the results of the above calculation (which was first made by Bunsen) do not agree at all with observed facts. According to what has just been said it would, under ordinary circumstances, be quite sufficient to wash a precipitate four times with a quantity of liquid ten times that of the solution adhering to the precipitate, whereas experience has taught us that a precipitate is far from being pure after being washed to this extent only. This disagreement arises from the false assumption that the amount of impurity on the filter is reduced to the mth part by mixing the precipitate with $(m-1)$ times the quantity of water and filtering off $(m-1)$ parts of liquid. That is not the case; for we have omitted to take into account in the above statement the phenomenon of *adsorption*, or surface condensation on solid bodies, *i.e.* the fact that in contact-surfaces between solids and liquids, the concentration of the dissolved substance is different from and always greater than in the rest of the solution. From this circumstance the amount of dissolved substance remaining with the precipitate after the latter has been drained is always more than would correspond to the quantity of adhering liquid, and, further, the amount which is removed by each washing is smaller than it would be on the above assumption. Both of these causes unite to render the washing of a precipitate less thorough than the foregoing theory would lead us to expect.

Almost nothing is known yet about the laws of adsorption. We can merely say that the amount of adsorbed substance is very probably proportional to the surface-area, and is also a function of the nature of the solid and dissolved bodies and of the concentration of the latter. With respect to this function, a knowledge of which would be requisite to the formulation of a rational theory of the washing of precipitates, nothing more can be said than that with a given nature and extent of surface, the amount adsorbed is most likely not quite proportional to the concentration, but diminishes more slowly than the latter.

To consider the behaviour of a precipitate upon washing, on the simple assumption that the amount of impurity adsorbed is proportional to the concentration of the solution,[1] let the ratio between the adsorbed

---

[1] The amount adsorbed cannot depend upon the absolute amount of the dissolved substance or of the solution, for, if we imagine the substance adsorbed by the solid to be in equilibrium with the solution, this equilibrium could not be destroyed by dividing the

amount x and the concentration of the solution c be k, so that the relation is

$$c = kx.$$

If now a quantity m of the washing liquid be added to the precipitate, which has originally adsorbed the amount x_0 of the dissolved substance, the residual quantity x_1 will be determined by

$$\frac{x_0 - x_1}{m} = kx_1,$$

seeing that the amount $(x_0 - x_1)$ has gone into solution, and has given with m of the solvent the concentration $\frac{x_0 - x_1}{m}$. If, when this solution has drained away thoroughly, a further amount m of solvent is poured on to the filter, the now residual portion x_2 of the adsorbed substance is given by the analogous equation—

$$\frac{x_1 - x_2}{m} = kx_2.$$

Eliminate $x_1 = \frac{x_0}{km + 1}$ from this, and we get

$$x_2 = \left(\frac{x_0}{km + 1}\right)^2,$$

and, generally, for n washings

$$x_2 = \left(\frac{1}{km + 1}\right)^n x_0.$$

This equation is similar in form to that given on solution into two parts at any given spot by a partition, and removing that portion of the solution which was outside the latter.

p. 17, excepting that here the amount of washing liquid m is multiplied by a coefficient k. In other words, the difference between the washing of a precipitate when we take adsorption into consideration and when we do not, is that in the latter case the action of the solvent is only partially accounted for.

As already stated, it is improbable that k could be taken as constant in the case of highly dilute solutions also. On the contrary, the presumption is that it diminishes rapidly with the diminution in the amount of adsorbed substance, thus further lessening the action of a prolonged washing. At any rate the difficulty which one experiences in actual practice in washing the last portions of an adsorbed substance out of a precipitate points to the coefficient k behaving in this way.

In the above calculation no account has been taken of the liquid retained in the pores of the precipitate by capillary attraction. It is easy to see, however, that if this were considered the resulting formula would be similar in structure to the one just given, although somewhat more complex; the residual portion of the impurity decreases continuously in a diminishing geometrical series with the number of the washings. Here, again, the rule holds good to add the wash-water in small quantities at a time, and to allow each portion to drain off thoroughly from the filter.

Adsorption effects are produced not merely by the precipitates themselves, but also by the filtering media, more especially by the cellulose of filter paper. In accordance with the end in view a finely porous texture is aimed at here, which is very favourable to the development of considerable adsorption actions. In an ordinary filtration we have only to consider the case of a small un-ribbed filter, which in washing is completely filled with water, in order that the edges too may be thoroughly cleaned. The above effects, however, become of consequence in those frequent cases in which a muddy liquid is only partially filtered through a dry filter, in order that some analytical estimation may be made in a given volume of the filtrate. *The first drops of the filtrate must therefore be discarded*, as they are much less concentrated than the rest of the liquid, in consequence of adsorption by the filter paper. The filter very soon reaches a state of equilibrium with regard to this adsorption, and the subsequent runnings possess the same degree of concentration as the original solution.

Alkaline liquids show this phenomenon in a marked degree; solutions of acids and neutral salts less.

7. *The Enlargement of the Crystalline Grains*

The fact, already mentioned on p. 13, that a fine powder of a crystalline precipitate is gradually converted into larger grains by digestion in the liquid, is of very wide application. The cause of this is to be sought for in the surface tension which exists on the boundary surfaces between solids and liquids, as on those between liquids and gases—the so-called free surfaces of liquids. This tension acts so that the surfaces in question are reduced in size, with the consequent enlargement of individual crystals (the total amount of precipitate remaining practically unaltered), *i.e.* with the coarsening of the grains.

This change is brought about through the smaller crystals of a precipitate being somewhat more soluble than the larger. It is true that this difference in solubility has not yet been proved experimentally, and, being so slight, it is hardly likely to be proved for some time to come. But, from the above-mentioned considerations with regard to surface tension, this conclusion follows of necessity from energy conditions. Through this difference in solubility of the larger and smaller crystals the liquid remains all the time supersaturated with regard to the former; the small crystals must therefore dissolve and the large ones grow.

But, it may be asked, how can this be explained in the case of insoluble substances? The answer is that there are none such. We have to go on the principle that *every substance is soluble;* the degree of solubility may vary greatly, but it can never be reduced to zero. It has in fact lately become possible not merely to prove but also to measure the solubility of such insoluble compounds as chloride, bromide and iodide of silver.

The rate at which this change progresses depends upon various circumstances. Thus it increases with the solubility of the substance; this is so marked that somewhat soluble precipitates like magnesium-ammonium phosphate usually come down coarsely crystalline at once, or at least become so in a very short time. Again the change proceeds more rapidly at a high temperature than at a low one. This depends on the one hand upon the increased solubility which most bodies show with rise of temperature, and on the other upon the much greater rate of diffusion of the dissolved substance, whereby its transport from the points of solution to those of deposition is facilitated.

The production of coarsely crystalline precipitates is to be aimed at not merely because these can be filtered off quickly, but also because they are purer and easier to wash than very fine ones. For the amount of impurity from adsorption is proportional to the surface, and is therefore smaller the coarser the grains are. But there is a limit here too, since large crystals readily enclose some of the mother liquor, which cannot of course be got rid of by washing. So far as is known, however, this never occurs in such crystalline precipitates as we obtain in actual analysis.

8. *Colloidal Precipitates*

Many amorphous bodies possess the property of an indeterminate solubility in water. The solutions which they form differ to some extent from ordinary solutions, standing between the latter and mechanical depositions or suspensions. From those solutions the substances in question are thrown down by various causes, such as warming, the addition of foreign bodies, and evaporation; many of them are thus made to lose the property of re-solution or re-suspension in water, while others still retain it. This capacity is, however, destroyed in probably every case by heating to redness.

Ferric oxide, alumina, silicic acid and most of the precipitated metallic sulphides belong to this class. They occur in analysis as gelatinous or flocculent precipitates, and are usually difficult to wash, since their fineness causes them to stop up the filter, while they also tend to pass through the latter after being washed for some time.

The tendency of substances to form colloidal or pseudo-solutions varies considerably; for analytical purposes it is best reduced to a minimum.

All substances of this kind are thrown down by

salt solutions; acids and bases act upon them in the same way, only more strongly as a rule—that is, of course, when these give rise to no chemical changes. The nature of the salt seems to exercise little influence, but the requisite concentration depends upon the nature of the colloid. If the salt solution is got rid of or diluted beyond a certain point, many of the precipitated colloids pass again into solution; others become changed by precipitation and are thus rendered insoluble. The latter is probably the more common, but the transformation takes place so slowly in many cases that it cannot be conveniently observed and made use of. Since salts, acids, or bases are usually present when such substances are thrown down in analysis, the latter generally appear as precipitates; when, however, the solution becomes diluted by washing, a point is reached at which a pseudo-solution may be reproduced. This begins in the upper layers of the precipitate. The resulting pseudo-solution meets in its passage through the remainder of the precipitate and the filter with more of the concentrated salt solution, and again undergoes precipitation in the pores, the latter being thus narrowed and the filter choked. The pseudo-solution ultimately penetrates the filter itself, or, as we say, the precipitate "goes through."

In order that this may be avoided we have to take care that a sufficiency of concentrated salt solution remains always mixed with the precipitate, and for this purpose a solution of some salt is to be used in washing instead of pure water itself. Since any salt produces the desired effect, we naturally choose one which can be afterwards easily got rid of, *i.e.* a volatile salt such as acetate of ammonium. If the solution has to be boiled, as in the separation of titanic acid ammonium acetate cannot be employed; instead of it sodium sulphate is to be taken.

In a few instances we obtain colloidal substances in analysis from solutions in which no salts are present, *e.g.* when a pure solution of arsenious acid is thrown down by sulphuretted hydrogen. Here no precipitate is formed, but a semi-transparent liquid which passes through a filter unchanged. To convert this into a filterable liquid, we must add either a salt or an acid to it, when the well-known yellow flakes separate sooner or later, according to the degree of concentration.

A second condition, which is favourable to the management of colloidal precipitates, is a somewhat high temperature. Many colloids separate completely when their pseudo-solutions are warmed; all of them change at higher temperatures into denser and less easily suspendable forms. Silicic acid, for example, becomes insoluble after prolonged drying over the water-bath, and alumina filters much more readily when digested for some hours in the liquid from which it has been precipitated.

The phenomena of adsorption are developed in a very high degree in the case of colloidal substances because of their extremely fine state of division, and retard the washing to such an extent that it can often not be finished within a reasonable time. This difficulty, however, is likewise lessened by all those conditions which go to render the precipitate more compact. In particular, adhering impurities are usually much more easily washed out after the ignition of the precipitate than before, the greatest degree of "condensation" being produced by this strong heating, sometimes indeed transformation into other, probably crystalline, forms. As a result of this condensation the surface is materially reduced, and, with that, the greater part of the adsorbed substance is liberated. A chemical transformation has a similar effect. Cobalt oxide, which has been thrown down by potash, cannot be washed free from the latter; this, however, is readily done from the metallic cobalt obtained by reducing the oxide with hydrogen. Due regard must always be paid in cases of this kind to any possible chemical interactions which may take place between precipitate and adsorbed substance upon ignition.

9. *Decantation*

Decantation offers a still simpler means of separating solids from liquids than filtration. Here the two substances are allowed to divide into layers, in virtue of the difference in their specific gravities, which is generally considerable, the (lighter) liquid layer being then poured off. It is not possible, however, to carry out a quantitative separation in this way, so that the procedure is merely applied in analysis as an aid to filtration, the liquid being run through a filter, which retains any particles of solid. The washing can be done in the same way, with a great saving of time, when dealing with very fine or colloidal precipitates which would readily stop up a filter. Substances which pass through a filter (such, for instance, as arsenious sulphide in pseudo-solution) do not deposit from the solution in which they are held; the reason for this is the same in both cases, and the same remedy applies to both.

Deposition can be hastened to a great extent by centrifugal force, which intensifies the separating pressure-differences in a marked degree.

10. *Separation of Liquids from Liquids*

The direct separation of one liquid from another can only be carried out in those cases where the liquids neither mix through nor dissolve in one another. Strictly speaking, of course, every liquid dissolves to some extent in every other, but this mutual solubility is often so slight that it may be practically disregarded.

Admixed liquors are separated by letting the heavier sink to the bottom of the containing vessel, which can also in certain cases be facilitated by centrifugal action; they are afterwards separated mechanically either by means of siphons, or, more conveniently, by a separating funnel. The smaller the dividing surface between the two liquids is, the more easily and thoroughly can the separation be effected.

This method of separation is applied in analysis in shaking up two non-miscible liquids together, when the object is to concentrate in one of these some substance which, while soluble in both, dissolves in them in different degree. A practically complete separation can only be achieved in this way by repeating the operation several times.

11. *Separation of Gases from Solids or Liquids*

On account of the great difference in specific gravity, gases and vapours separate very easily and quickly from solids and liquids, so that much use is made of this procedure. Since, however, only a comparatively small number of substances are gaseous at the ordinary temperature, the separation is generally carried out at a higher one, *i.e.* either under distillation or sublimation.

In this latter case the process is particularly easy of control, seeing that the substance in question merely becomes gaseous for the moment, relapsing back on condensation to the liquid or solid state. In this way the large amount of space required to hold any considerable amount of gas is dispensed with; and, since the condensation of the vapour is induced in an apparatus designed for the purpose, a very convenient and almost complete separation is effected. The only part of the vapour which remains unseparated is that which fills the distilling vessel at the close of the operation, and even this can be expelled by leading in some other suitable gas or vapour.

12. *Separation of Gases from one another*

Since all gases—in so far as they do not affect one another chemically—are miscible in any proportion, the direct separation of a mixture of gases into its constituents cannot be carried out by mechanical means. A partial separation can be made by *diffusion*, seeing that light gases diffuse through others or through any porous material more quickly than heavy do. But it is impossible to effect a complete separation in this way, and consequently the method has been applied more for the purpose of proving the presence of different constituents in a mixture of gases than of separating them from one another.

The use of porous partitions often permits of much better separations than free diffusion into another indifferent gas, especially as the process can be repeated. There are permeable materials which appear to render a complete separation possible in the case of individual gases; for instance, hydrogen passes readily through heated palladium or red-hot platinum, while all other gases are stopped. It appears, however, to be tolerably certain that we have here not merely a simple sieve-effect, but that there is an intermediate action of the nature of solution or chemical combination.

The different extent to which different gases are absorbed by porous solids may also furnish another method of partial separation. Practical use is made of this property in the removal of bad-smelling gases from the air by freshly ignited charcoal, although it has hardly been applied so far to analytical purposes.

All these methods of gas-separation are very imperfect. Consequently, when a complete separation is required, one of the gases has to be transformed— usually by some chemical change—into a liquid or solid, which can then be isolated from the other gaseous constituents without any trouble.

CHAPTER III

PHYSICAL METHODS OF SEPARATION

1. *General Considerations*

AFTER one has considered the possible cases in this problem of separating several substances, and has examined these with respect to their practicability, it is seen that there are certain instances (*e.g.* among miscible liquids and gases) which do not allow of a separation, or, at least, of one sufficiently exact. In other cases, again, experimental or other difficulties occur. It is therefore necessary here to change the material under examination into some other state, in which the wished-for separation can be easily and completely carried through. Those latter (favourable) cases consist in solid-liquid, liquid-gaseous, or solid-gaseous combinations, into which the original ones must be brought by some suitable procedure.

Alterations of temperature and the use of solvents are the means by which such changes can be effected. The first of these is mostly applicable to those cases in which one of the constituents can be vaporised. For the direct separation of a liquid from a solid can never be complete, on account of the residue of liquid which always remains behind moistening the solid; and subsequent washing is not possible here, as when separating solutions from precipitates. On the other hand, the use of solvents is preferred for the separation of solids from liquids. And, since there are many more substances capable of solution than of vaporisation, the latter procedure is followed to much the greater extent.

Thus the aids to bringing about suitable conditions for separation are restricted for the most part to the two methods of distillation and solution. If those should not be feasible, chemical means must be brought to bear, and these will be considered later on.

2. *The Theory of Distillation*

Speaking generally, every solid or liquid can be converted into vapour at any temperature, but this change may be said only to take place in a measurable degree in the case of certain substances (not in all), and above certain limits of temperature, which are different for different bodies. The law of this transition is simple and applicable all round. It is this—conversion into gas or vaporisation takes place so long as the gas or vapour has not reached a certain degree of concentration at the surface of the vaporising substance, this concentration being dependent only upon the nature of the latter and upon the temperature; and, without any exception, this characteristic concentration increases with rise of temperature.

The above law is usually put in this way—that, corresponding to each temperature, there is a particular vapour pressure. In the case, therefore, of other gases being present, the partial pressure of the vapour in question must be given. But the only way of getting at this is to determine the proportion between the quantity of this vapour and that of the other gases, and, after dividing these by the corresponding specific gravities, to distribute the total pressure in the ratio of the numbers thus found. The definition which has been given, however, possesses the advantage over this of greater simplicity, since the measurement of the concentration involves merely a knowledge of the amount and the volume; certain abstract difficulties with respect to the partial pressure are also avoided by it.

It must be again emphasised that in this law we have merely to do with the concentration, *i.e.* the partial pressure of the vapour itself. Whether other gases or vapours are present in the same area has no influence on the equilibrium (or at least only a secondary influence, which need not be referred to here).

Distillation is thus a very simple matter in the case of substances whose vapour pressures differ widely, *i.e.*, practically speaking, in the separation of volatile from non-volatile bodies. The temperature of the mixture is raised to the boiling point of the volatile constituent, or, in other words, to the temperature at which its vapour pressure just exceeds that of the atmosphere, and the volatile portion is condensed by cooling if wanted. At the end of the operation the distilling vessel remains filled with the vapour of the volatile constituent; but the latter can be displaced by a current of some indifferent gas, the apparatus being constructed with this end in view.

Since the boiling point depends upon the external pressure, it can be lowered, if desired, by reducing the latter. For this purpose the distilling apparatus must be constructed air-tight, and the air pumped out of it before beginning to distil. Seeing that we have only to do with partial pressure here, we can achieve the same end by the admixture of another gas or vapour, *i.e.* by distilling in a current of gas or vapour. Which of those two should be chosen will depend on the circumstances of the case. If the distillate is to be collected, a current of vapour is preferable, since a mixture of vapours can always be condensed without loss, while an admixed gas carries away with it a quantity of the volatile substance corresponding to its own volume, and to the vapour pressure of the latter at the temperature of the condenser. But, should the distillate not be wanted, it is often more convenient to distil in a current of gas.

The amount of the volatile vapour carried off by the current of gas is—as just stated—proportional to its vapour pressure at the temperature of distillation, and to the volume of the admixed gas. If B is the barometric pressure and p the vapour pressure at the temperature in question, the volumes v and V of the two ingredients of the gaseous mixture stand in the ratio of the partial pressure $\frac{p}{B-p}$, and the volume of vapour v of the distilling substance at the atmospheric pressure B is $v = V\frac{p}{B-p}$; on multiplying this value by the vapour density, the weight of the distillate is obtained.

3. *Distillation of Mixed Liquids*

Two liquids may either not mix at all, or they may be miscible partially or completely and in any proportion. The first-mentioned case is to be regarded as the theoretically impossible limit, which, however, is often nearly enough approximated to for practical purposes. The theory of the distillation of non-miscible volatile liquids has just been given, under distillation in a current of gas, of which it is merely a particular case. But the temperature of volatilisation here, unlike that above, is no longer to be chosen at will, but is conditioned by the boiling point of each of the two liquids at the partial pressures of their vapours; both of those boiling points are necessarily identical, and lie below the boiling temperature of the more volatile ingredient, seeing that its partial pressure is of necessity smaller than the total pressure, or than that of the atmosphere.

The quantitative proportion between the two substances under distillation remains constant in this case, so long as both vapours are present in the escaping bubbles. If p_1 and p_2 represent the two partial pressures, and d_1 and d_2 the two vapour densities, then the relative weights m_1 and m_2 are in the ratio—

$$\frac{m_1}{m_2} = \frac{p_1 d_1}{p_2 d_2}.$$

The same laws hold good if one of the two substances is a solid, and insoluble in the other.

The two substances are, of course, readily separable in the common distillate, seeing that we are going on the assumption of their being non-miscible. Where, then, would be the object of such a distillation? As a matter of fact it is only made use of in place of distillation in a current of gas or vapour (which has been already explained), in order to separate some volatile substance from others which are non-volatile.

If the two liquids are partially miscible, in the sense

that some of the first dissolves in the second and *vice versâ*, but that the two solutions as a whole do not mix, the above laws still hold good in part. In the first place, emphasis has to be laid on the fact that both solutions have the same vapour pressure, seeing that they are both made up of the same constituents. One, therefore, obtains a constant mixture of the two substances on distilling, so long as two layers are present in the retort, and this mixture again separates in the receiver into two non-miscible, mutually saturated solutions. It is thus impossible to effect any further separation by such a distillation; the case, therefore, does not concern us in analysis.

But if the two non-miscible portions, A with some of B and B with some of A, are each distilled alone, a further separation can of course be brought about. This comes, however, under the distillation of homogeneous solutions, to which we shall now refer.

In homogeneous mixtures of volatile substances, the combined vapour pressure is always lower than the sum of the vapour pressures of the constituents at the same temperature, the vapour pressure of any volatile body being always diminished by the solution of some other in it.

The behaviour of homogeneous mixtures on distillation is most easily understood by making a graphic representation of the combined vapour pressure as a function of the composition. If, in the subjoined figure, the ordinate aa represents the vapour

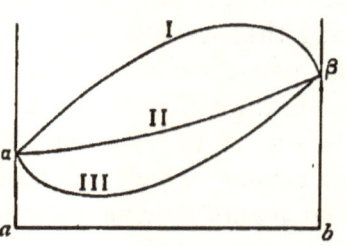

pressure of the first liquid, and the ordinate $b\beta$ that of the second, then the vapour pressures of all possible mixtures of both, drawn according to their percentages between a and b, will form a continuous curve corresponding to one of the three types I, II, and III. In other words, we shall either have a mixture whose vapour pressure is higher than that of all the others (curve I), or lower than these (curve III), or, finally, mixtures with vapour pressures between those of the several liquids, without any maximum and minimum (curve II). In the first type (I) the boiling points of the mixtures lie below the mean value, and we get the mixture of lowest possible boiling point; in III we have the mixture of highest possible boiling point; while II includes all the boiling points of mixtures which lie between those of the two constituents. Now the law holds *that, in the case of the mixture which possesses the highest or the lowest boiling point, the vapour must have the same composition as the liquid itself.* Such mixtures consequently behave as homogeneous liquids, and they cannot be separated by distillation. It follows therefore that liquids belonging to types I and III can only be resolved by distillation into (1) the particular mixture of highest or lowest boiling point, and (2) the liquid which is present in excess with respect to that mixture, any further separation being impossible in this way.[1]

In case II, however, a more or less perfect separation can be effected. If any given mixture of such liquids be raised to boiling, the ratio between the two

---

[1] Such mixtures of constant boiling point (*e.g.* of hydrochloric acid and water) have often been regarded as chemical compounds, but wrongly so. The fact that they are not is proved by their compositions altering continuously with alteration of the pressure under which they are distilled.

constituents in the vapour will be different from that in the liquid; a larger proportion of the more volatile ingredient will generally be present in the former. The same holds good with regard to the distillate, the separation being more complete the greater the difference between the vapour densities of the two liquids. By re-distilling the first portion of the distillate, a further separation is effected, so that in time we get the more volatile liquid collected in the distillates, while the less volatile remains behind in the residues.

Repeated distillations like this can be carried out automatically by partially condensing the vapour, and compelling each succeeding portion to pass through that which has just been liquefied. A great many different kinds of distilling apparatus have been constructed with this end in view, which, however, it would be out of place to describe here. These effect an approximate *fractional distillation*, but cannot be used for quantitative purposes. When a quantitative separation is required, the only method is that of chemical transformation, whereby one of the constituents is changed into a solid or a non-volatile state; and this applies also to the separation of mixtures of constant boiling point (p. 37).

4. *Separation by Solution*

Homogeneous mixtures of different substances are termed solutions. For our purposes here we need only consider *liquid* solutions, which are formed when substances of any state of aggregation are brought into contact with suitable liquids. According to the state of aggregation of the substance to be dissolved, we distinguish between the three cases of solution of gases, of liquids, and of solids in liquids. The process of separation by solution consists, generally speaking, in finding and using a liquid which readily and abundantly dissolves one of the substances to be separated, while it leaves the other practically unaffected. The latter condition cannot be absolutely fulfilled, but there are numberless cases where it is approximated to with sufficient accuracy. By the use of such a solvent the two substances in question are rendered separable by mechanical means; in the case of solution of liquids in liquids it is necessary, in addition, that these shall not be miscible with one another.

5. *Solutions of Gases*

The law according to which gases dissolve in liquids is that, given a state of equilibrium, the concentration of the gas bears a constant relation to that of the solution. By concentration is meant, of course, the amount present in unit of volume. Since the concentration of a gas is proportional to the pressure, so is also the quantity of it dissolved. The ratio depends upon the nature of the substance and the temperature. All gases dissolve in water, alcohol, and similar liquids, albeit in most cases to only a slight extent. In order to produce a solution in a state of equilibrium, *i.e.* a *saturated* solution, it is needful to bring the largest possible surface of gas and liquid into contact, and to disseminate the dissolved substance throughout the liquid, this being effected by passing the gas in very small bubbles, shaking the liquid, and other mechanical means.

In most cases what we want is not so much a saturated solution, as to attain to the most perfect absorption practicable. Here, however, the difficulty meets us that, in our case of a gaseous mixture, the concentration (or the partial pressure) of the gas under absorption becomes less, the further the separation by solution has proceeded. We must therefore apply the principle of *counter-currents* by letting the mixed gases and the solvent liquids meet one another travelling in opposite directions. In this way the almost saturated solution meets with fresh gas, while the new liquid comes first into contact with the gaseous mixture from which nearly all the soluble constituent has been already removed. We thus ensure, in the former case, the greatest possible saturation and therefore the smallest expenditure of solvent, and, in the latter, the completest possible absorption of the last traces of soluble gas.

For quantitative purposes the simple absorption of gases in liquids can be applied but seldom, because the absorption-coefficients of most gases are too small and approximate too closely to one another to allow of this. A few gases, however, such as the hydrogen-halogen acids can be separated well enough in this way from hydrogen, nitrogen, air, etc. In most cases a chemical transformation is necessary in the first instance; even in that of the hydrogen-halogen acids, just mentioned, there is reason for believing that chemical interactions take place on their solution in water.

When separating gases by absorption it is necessary to bear in mind that the unabsorbed gas carries away with it an amount of the solvent corresponding to the vapour pressure, and this must either be allowed for, or it must be prevented from escaping by some suitable means.

6. *The Drying of Gases*

A very frequently occurring case in gas-separation is the drying of gases, *i.e.* the abstraction from them of any water vapour that may be present. For this purpose either liquid solvents such as strong sulphuric acid, or solid absorptives like calcium chloride, caustic potash, and phosphorus pentoxide are employed. In order to carry out the operation successfully, we have to bear in mind the considerations that have just been mentioned above. For example, it is much more efficacious to spread the strong sulphuric acid over some porous material like pumice stone, and thus to establish a large acting surface, than simply to let the gases bubble through the liquid acid. In this case, also, most of the interactions that ensue are of a chemical nature.

7. *Two Non-Miscible Liquids; the Theory of the Extraction of a Dissolved Substance from one Solvent by shaking this up with another*

When two non-miscible liquids, contained in the same vessel, are brought into contact with a third substance which is soluble in both of them, the latter so distributes itself that its concentration in the one liquid bears a constant relation to that in the other. This law was discovered by Berthelot and Jungfleisch, and has since been frequently verified experimentally. Under certain conditions it seems to undergo modification, but this will be explained later on. The simple statement just given is sufficient for a survey of the subject in its main points.

If x_0 represents the quantity of dissolved substance present in amount l of the first solvent, then when this solution is shaken up with amount m of the second solvent, x_1 remains behind in the first and $x_0 - x_1$ passes into the second. The quantity x_1 is got by the equation

$$\frac{x_1}{l} = k\frac{x_0 - x_1}{m}, \text{ or } x_1 = x_0\frac{kl}{m + kl},$$

since $\frac{x_1}{l}$ and $\frac{x_0 - x_1}{m}$ represent the two concentrations, and k the constant ratio-number between these, or the distribution coefficient.

A second shaking up with a similar quantity m of the second solvent gives

$$\frac{x_2}{l} = k\frac{x_1 - x_2}{m};$$

or, after substituting for x_1 its equivalent given above,

$$x_2 = x_0\left(\frac{kl}{m + kl}\right)^2,$$

and for n shakings

$$x_n = x_0\left(\frac{kl}{m + kl}\right)^n.$$

Here, again, the form of equation is the same as in the theory of washing precipitates, and a similar conclusion is to be drawn here also, viz. that with a given amount of solvent a more perfect separation is effected when the shaking up is done with a great many small portions, rather than with a few large ones. For the rest, the result depends upon the value of the distribution coefficient k; the smaller this is, $i.e.$ the smaller the ratio of concentration between the first and second solvents, the quicker is the progress of the operation. But it is as impossible to effect an absolute separation by shaking up as it is to wash a precipitate completely.

8. *Solutions of Solids*

The law of solubility for solids is that equilibrium or saturation ensues upon a definite concentration of the solution; the value of this concentration depends upon the nature of the substances and upon the temperature, the concentration usually increasing with rise of temperature, although in certain cases it falls.

The value of this saturation-concentration at a given temperature is entirely dependent upon the state of aggregation of the solid with which the solution is in contact, altering as this alters. We have to note especially here that a definite and in every case different solubility attaches to the various polymorphic and allotropic forms, and the various hydrates, etc., of one and the same substance. For this reason such an expression as "the solubility of sulphur" is quite indefinite, even although the solvent and the temperature are stated; to make it complete, the particular modification of sulphur in question must be given.

It has already been mentioned that amorphous substances have, as a rule, no definite solubility, so far at least as they form colloidal solutions. They are, however, always much more soluble than the corresponding crystalline compounds. Most crystalline precipitates appear to come down amorphous in the first instance, and then to change more or less rapidly into the crystalline state, this being especially noticeable in the case of carbonate of lime. On account of the greater solubility of the amorphous modifications, this change into the crystalline form must always be awaited when separations are in view. The mode by which it can be hastened has been already described in detail on p. 22.

The separation of two solids, by means of a solvent in which only one of them is soluble, is subject to essentially the same laws as the washing out of precipitates (see p. 27). The actual treatment with the solvent is most conveniently effected in some vessel and not on the filter. The resulting solution is then poured through the filter, and the residue in the vessel treated with fresh quantities of liquid until one is certain that everything which is soluble has dissolved, when the residual solid is finally thrown on to the filter and washed. The reason for proceeding in this way is that it is by no means easy to ensure that every particle of solid on a filter shall be thoroughly exposed to the action of the solvent.

Since the rate of solution is proportional to the extent of contact surface, it is always advisable to reduce the solid to fine powder, should it not be in that state already; this applies especially to sparingly soluble substances. Warming is also to be recommended in most cases.

If it should be desired to carry through the separation with the least possible quantity of solvent, the latter may be distilled off from the extract—assuming the dissolved substance to be non-volatile—and used over again. These operations repeat themselves automatically in an *extraction apparatus*, which consists of a distilling flask with reflux condenser attached; between these the filter containing the substance to be extracted is placed, this allowing the recondensed liquid to percolate it time after time. There have been a good many different forms of apparatus devised for the above purpose, the most practical being those in which a self-acting siphon returns the solvent—which after percolation lodges in the interstices round the filter—back to the distilling flask.

9. *Several Soluble Substances*

Speaking generally, every substance must be looked upon as soluble, and thus a separation by means of a solvent is of necessity always imperfect, seeing that a small portion of the "insoluble" ingredient is invariably dissolved; but should the latter be so minute as to be negligible, the substance in question is considered insoluble. In cases where this solubility is appreciable and has to be taken into account, it is important to know the laws of the collective solubility of several substances. For sparingly soluble bodies, which undergo no change in dissolving, the law holds that these dissolve independently of one another, until the saturation-concentration is arrived at for each one. It is the same law which applies to the simultaneous solution of several gases in a liquid, and also to the common vapour pressure of non-miscible liquids.

In this simple form, however, the law of independent solubility finds but little application. In cases where salts—or, generally, electrolytes—are concerned, the solubilities exert a mutual influence if the different substances contain a common ion. These relations will be discussed later on.

CHAPTER IV

CHEMICAL SEPARATION

§ 1.—THE THEORY OF SOLUTION

1. *Introduction*

IF a separation can neither be effected directly nor by physical means, we have the most frequently occurring case of all, viz. the transformation of the substance to be separated into some other chemical compound whose condition admits of a mechanical separation. Here also, as before, we have to aim at getting mixtures of solid and liquid on the one hand, or solid and gas or liquid and gas on the other.

In order to fully appreciate the processes which go on in solutions of most substances, it is necessary that we should first discuss the theory of solution and the state of a substance after it has been dissolved. The recent developments in this branch of the subject have caused the theory of analytical reactions to enter upon an entirely new phase—indeed it is only through them that this theory has become a truly scientific one; the progress of analytical chemistry centres mainly in this point.

2. *The State of Substances in Solution*

The idea frequently given expression to by earlier investigators — that in dilute solution substances assume a condition which is similar to the gaseous state—has now become a strictly scientific theory, thanks to the pioneering labours of van't Hoff. In preceding sections of this book stress has been repeatedly laid upon the agreement between the empirical laws which have been worked out for dissolved substances on the one hand and for gaseous on the other, with respect to solution and saturation; this agreement extends so far that matter in the two above states obeys the same law with the same constants, with only this difference, that in place of ordinary gaseous pressure, *osmotic pressure* comes in for dissolved substances. Osmotic pressure means the pressure which is exerted on a boundary surface separating a solution from the pure solvent, when there is at this surface a septum which permits only the solvent and not the dissolved substance to pass through.[1]

Just as vapour density determinations at different temperatures and pressures have given us information with regard to the molecular complexity of substances in the state of vapour, so has the study of solutions shown that the formulæ usually ascribed to many substances do not apply to them when in aqueous solution; they must in fact have when in solution a smaller molecular weight than that corresponding to

[1] For fuller details see the author's *Outlines of General Chemistry* (English translation by James Walker), p. 129; or his *Lehrbuch der allgemeinen Chemie* (2nd ed.), vol. i. p. 651; or M. M. P. Muir's translation of that portion of the latter which treats of *Solutions*.

the lowest possible formula. The interpretation of this problem was at first attended with great difficulties, which were, however, ultimately surmounted by Arrhenius in his *Theory of Electrolytic Dissociation.* This investigator perceived that the deviations just referred to occur only in the case of substances which behave as *electrolytes,* and he was enabled to formulate the laws of electrolytic conductivity and to explain the deviations of the solutions in question by the assumption that *salts do not exist as such in aqueous solution, but are dissociated more or less completely into their constituents or ions.*

It would be out of place to enter in detail into the various methods by which this assumption has been confirmed and justified; it will be taken for granted here, but a number of new proofs of its applicability will be brought forward in the course of the following pages.

3. *Ions*

It was early forced upon the observation of chemists that salts were binary compounds. Berzelius looked upon them as being made up of acid and base, or rather of acid anhydride and metallic oxide; this, however, placed him in the difficulty of having to regard halogen salts as constituted differently from oxygen ones, although there is nothing in the behaviour of the two classes which demands or justifies such a distinction. Liebig, together with a number of other chemists, subsequently perceived that the true constituents of salts are the metal on the one hand and the halogen or acid radicle (*i.e.* salt minus metal) on the other, and Faraday gave to those constituents the name of *ions.* We have to distinguish between positive ions or cations (metals and complexes which behave like metals, *e.g.* NH_4), and negative ions or anions (halogens and acid radicles like NO_3, SO_4, etc.).

In aqueous solutions of electrolytes the ions are usually partly combined and partly free. In the case of neutral salts the uncombined portion is by far the greater, and it increases as the solution becomes more dilute. The properties of a dilute salt solution are thus dependent upon those of its free ions rather than on the properties of the dissolved salt as such, or of the ions which are combined. The analytical chemistry of salts thus becomes enormously simplified under this law; what we have to establish is not the analytical properties of the possible salts, but only those of their ions. Supposing we made the assumption that fifty anions and fifty cathions were to be considered, these might form 2500 possible salts; and, if each salt showed some individual reaction or reactions, the behaviour of 2500 different substances would have to be ascertained. But, since the properties of salts in solution are merely the sum of the properties of their ions, it follows that we only require to know $50 + 50 = 100$, in order to master the whole 2500 possible cases. As a matter of fact, analytical chemistry has long made use of this simplification; we know very well, for instance, that the reactions of the salts of copper are the same with respect to the copper they contain, whether it be the sulphate, nitrate, or any other salt that is examined. But it has been left for the electrolytic dissociation theory to formulate this relation scientifically.

And the above theory not only explains the great simplicity of the analytical system in this way, but also those complications which we find by experience to occur in particular cases. While the numerous metallic chlorides all give the reaction of chlorine with silver, other chlorine compounds, such as potassic chlorate, the salts of the chloro-acetic acids, chloroform, etc., do not. Chloroform may, however, be at once eliminated here, for it is not a salt, and cannot therefore show the ion-reactions. Now the reason why the salts just named do not give the characteristic chlorine reaction, although they are salts and contain chlorine, is because they contain no chlorine ions. The ions of chlorate of potassium are K and ClO_3; the salt thus gives the reactions of the potassium ion and of the ion ClO_3, and no other reactions are to be looked for. In all cases, therefore, in which an element is a constituent of a compound ion, it loses its ordinary reactions, and in place of these we get new ones characteristic of the latter.

How the ionic condition is recognised and how the degree of dissociation is measured, are questions which cannot be discussed in detail here. But it may just be mentioned that electrolytic dissociation and electrolytic conductivity run on parallel lines, and that conclusions can be drawn with respect to the former by measuring the amount of the latter under given conditions. Besides this there are numerous other auxiliary aids, the application of which has led to the same results as the electrolytic conductivity.

The further question—which are the ions of any given salt—is not always easy to answer. Thus, it used to be supposed that potassium platinichloride was a chlorine compound of the same nature as metallic chlorides generally, whereas we now know that its ions are 2K and $PtCl_6$, and that it is therefore the potassium salt of hydro-platinichloric acid. As a consequence, it gives no chloride of silver with silver nitrate, but a yellow precipitate of silver platinichloride, Ag_2PtCl_6. We are able to decide the point in question chemically, by observing which are the complexes that interchange themselves with the ions of other salts. The electrolysis of the salts also furnishes an independent proof, since the cathions move in the direction of the positive current and the anions in that of the negative. Thus Hittorf, in electrolysing sodium platinichloride, found that the platinum and chlorine together betook themselves to the anode, while the potassium went to the cathode.

The parallelism between electrolytic conductivity and capacity for chemical interaction is one of the most important aids to judging of the ionic condition. Both properties run alongside of one another throughout, so that Hittorf has given the following definition :— *Electrolytes are salts*, *i.e.* binary compounds, which are capable of exchanging their constituents instantaneously. Since, for the purposes of chemical analysis, those reactions are of most importance which go on at the greatest possible rate, the ones referred to here are practically all ion-reactions.

4. *The Varieties of Ions*

In accordance with the fact of salts being binary compounds, the ions fall in the first instance into two classes, named by Faraday cathions and anions. The former move in the direction of the positive current when a stream of electricity is passed through an electrolyte, *i.e.* through a compound containing ions, and we therefore assume that they are combined with amounts of positive electricity which, according to

Faraday's law, are equal for equivalent quantities of different ions. The anions move in the opposite direction, and are therefore combined with negative electricity, the amounts of which are likewise equal for equivalent quantities of different ions. The word "equivalent" is used to denote those quantities of opposite ions which unite to form a neutral compound. These contain equal amounts of electricity, but of opposite sign; for, in an electrically neutral liquid, the sum total of the quantities of positive electricity must be equal to the sum total of the negative.

Since ions in solutions comport themselves as independent substances, it has been found possible to determine their molecular weights. This has resulted in showing that we have to distinguish between mono- and polyvalent ions, in accordance with what is demanded by the molecular weight determinations of non-dissociated compounds. The ions of potassic sulphate, K_2SO_4, for example, are $2K$ and SO_4; it follows from what has been said above that an electrically and chemically neutral solution of this salt must contain an amount of electricity for each SO_4-ion equal to that of the two K-ions, *i.e.* the ion SO_4 is charged to double the extent with negative electricity that the ion K is with positive. In the same way it follows from the formula $BaCl_2$ that the ion Ba must be divalent with respect to Cl. The following are the most important ions :—

A. CATHIONS

(*a*) Monovalent: H (in the acids), K, Na, Li, Cs, Rb, Tl, Ag, NH_4, NH_3R to NR_4 (R being an

organic radicle), Cu (in cuprous compounds), Hg (in mercurous compounds), etc.
(b) Divalent: Ca, Sr, Ba, Mg, Fe (in ferrous salts), Cu (in cupric salts), Pb, Hg (in mercuric salts), Co, Ni, Zn, Cd, etc.
(c) Trivalent: Al, Bi, Sb, Fe (in ferric salts), and most of the rarer metals of the earths.
(d) Tetravalent: Sn (doubtful), Zr.
(e) Pentavalent: None known for certain.

B. Anions

(a) Monovalent: OH (in bases), F, Cl, Br, I, NO_3, ClO_3, ClO_4, BrO_3, MnO_4 (in permanganates), and the anions of all the other monobasic acids, *i.e.* acid molecule minus an atom of hydrogen.
(b) Divalent: S, Se, Te (?), SO_4, SeO_4, MnO_4 (in manganates), and the anions of all the other dibasic acids.
(c) Tri- to hexavalent: The anions of the tri- to hexabasic acids. Elementary anions with a valency of more than two are unknown.

When it shall be necessary or desirable to designate the ions as such in the subsequent portion of this book, the formula of the cations will be accented by a dot (˙) and that of the anions by a dash (′). K˙ thus represents the potassium ion as present, for example, in an aqueous solution of potassic chloride, while the corresponding chlorine ion is denoted by the symbol Cl′. Polyvalent ions are marked with as many dots or dashes as correspond to their valencies or electric charges.

5. *Some further Details*

A few details with respect to the degree of electric dissociation or ionisation of the more important compounds are necessary for a proper review of analytical reactions, and they shall therefore be given here.

Non-electrolytes comprise the organic compounds with the exception of the typical acids, bases, and salts; also solutions of substances in solvents like benzene, carbon disulphide, ether, and similar liquids. Solutions in alcohol constitute a transition to the electrolytes, seeing that in them an ionisation of salt occurs, although as a rule only in very slight degree. The substances and solutions named at the beginning of this paragraph are not in truth to be looked upon as absolutely undissociable, just as there is no such thing as an absolute non-conductor; the line can only be drawn here, as in analogous cases, where observation and refined measurement reach their limits.

Electrolytes comprise salts in aqueous solution, the term "salt" including here and in the following pages both acids and bases, acids being salts of hydrogen, and bases salts of hydroxyl. Solutions of salts in alcohols are likewise ionisable, although to a far less extent; the ionisation is greatest in the case of methyl alcohol, and diminishes (for the same substance) with increase in the molecular weight of the alcohol.

Of the different kinds of salts, neutral ones are the most strongly ionised; aqueous solutions of medium concentration (such as are generally employed in analyses) usually contain much more than half the total salt in the condition of free ions. And salts differ in this respect among one another in so far that those with monovalent ions, such as KCl, $AgNO_3$, and NH_4Br, are the most dissociated, the dissociation decreasing as the valency of the ions rises. The nature of the metal and of the acid radicle has otherwise but little influence on the degree of ionisation of a salt. A few exceptions must be noted here. The halogen compounds of mercury are ionised very slightly, those of cadmium a little more, while those of zinc stand more or less midway between them and salts generally; again, of the halides, the iodine compounds show the smallest, and the chlorine compounds the greatest ionisation.

A much greater diversity exists among the acids and bases. These show degrees of ionisation corresponding to what is somewhat loosely termed their "strength," *i.e.* the strongest acids and bases are the most completely ionised.

The *strong acids*, whose ionisation is of the same order as that of neutral salts, are the halogen hydrides (with the exception of hydrofluoric acid, which is only moderately ionised), nitric, chloric, perchloric, sulphuric, and the polythionic acids.

Moderately strong acids comprise phosphoric, sulphurous, and acetic acids, which are not ionised beyond about 10 per cent under ordinary conditions.

The *weak acids*, with an ionisation of less than 1 per cent, are carbonic acid, sulphuretted hydrogen, hydrocyanic, silicic, and boracic acids. The ionisation of the two last is so slight as to be scarcely measurable.

Under the *strong bases* we have the hydroxides of the alkali and alkaline earth metals and of thallium, and also the organic quaternary ammonium compounds.

All these are ionisable to about the same extent as neutral salts.

The *moderately strong bases* are ammonia and the amine bases of the fatty series, oxide of silver, and magnesia.

The *weak bases* comprise the hydroxides of the di- and trivalent metals with the exception of those named above, the amine bases of the aromatic series (when the nitrogen is linked to the aromatic nucleus), and the greater number of the alkaloids.

Further details on the subject will be given in the special sections of this book where necessary. It is very important that we should fix the large groups—just mentioned—in our memories, since a critical review of analytical reactions is directly dependent to a large extent upon the conditions which have just been explained.

§ 2. CHEMICAL EQUILIBRIUM

6. *The Law of Mass-action*

In discussing chemical equilibrium we have two cases to take into consideration, viz. homogeneous and heterogeneous equilibrium. Homogeneous equilibrium exists in those spaces in which there is never a surface of separation, *i.e.* in gases and homogeneous liquids. As a matter of principle homogeneous solids must not be excluded, but, practically speaking, they do not come into question.

The law of homogeneous equilibrium may be stated thus: If we have a reversible chemical reaction corresponding to the general equation

$$m_1A_1 + m_2A_2 + m_3A_3 + \ldots \rightleftarrows n_1B_1 + n_2B_2 + n_3B_3 \ldots$$

in which the symbol $\rightleftarrows$ is used to indicate that the interaction may go on equally well from left to right (of formula) as from right to left; and if a_1, a_2, a_3 . . . and β_1, β_2, β_3 . . . represent the concentrations of the substances A_1, A_2, A_3 . . . and B_1, B_2, B_3 . . . ., while m_1, m_2, m_3 . . . and n_1, n_2, n_3 . . . stand for the respective numbers of the molecules taking part in the reaction, then the following equation holds good—

$$a_1{}^{m_1} a_2{}^{m_2} a_3{}^{m_3} \ldots = k \beta_1{}^{n_1} \beta_2{}^{n_2} \beta_3{}^{n_3} \ldots,$$

k being a coefficient depending upon the nature of the substance and the temperature.

This law is of very general application. If the amount of the particular substance divided by the total volume is taken as the concentration, it must be looked upon as an approximate law, which only becomes true in the limit-case, *i.e.* for dilute solutions. With a suitable definition of concentration it can be made generally applicable, but there is no such definition known yet for concentrated solutions. The simple definition just given is amply sufficient for our purposes.

All those substances are to be considered as taking part in the reaction which undergo a change, and which alter in their concentration. There are, it is true, special cases in which the first of these two conditions is fulfilled, but not the second. This occurs more particularly when the reaction goes on in a solution and the solvent takes part in it. In such a case the respective factor a^m or β^n remains constant, and can be incorporated with the coefficient k.

When ions share in the reaction, they are to be considered as being separate substances. The ordinary formulæ must therefore not be given in an equation to the electrolytically dissociated bodies, but their state of dissociation must be denoted. Thus potassic chloride in very dilute solution—when the salt is dissociated completely—could not be written as KCl in such a formula, but rather as $K^{\cdot} + Cl'$ (the dot and dash indicating the ionic condition, the former standing for cathions and the latter for anions). In the cases which have yet to be explained, examples will be given of this kind of formula.

The above law, now expressed mathematically, is nothing else than the most general form of the law of mass-action which was propounded by Wenzel more than a hundred years ago, and according to which the chemical action of any substance is proportional to its acting mass or concentration. The law may now be considered as being of general application, seeing that it has been abundantly verified in a great variety of ways, more especially within the last few years. Certain exceptions, which for a time seemed to stand in its way, have been brought into thorough conformity with it by the dissociation theory and the consequent treatment of ions as separate chemical individuals. The theory of electrolytic dissociation or ionisation has thus been instrumental here also in filling up what has hitherto been a gap in the system of theoretical chemistry.

The one limitation to which the ions are subject in virtue of their freedom is that positive and negative ions must always be present in equivalent quantities throughout. This limitation does not need to be expressly specified in the formulæ; they merely give a relation between the concentrations of the different ions, and this usually finds its expression in the coefficients of those concentrations.

7. Applications

The application of the theory of homogeneous equilibrium to the condition of dissolved electrolytes is one of the most important to which it has been put. In solutions of this kind a state of equilibrium subsists between the ions of the electrolyte and the non-dissociated portion, which can be expressed by the above formula; the correctness of the formula in its widest extent has also been proved and corroborated by the independent measurement of this condition.

To take the simplest possible case, if we have a binary electrolyte C, which can break up into the ions $A^{\bullet}$ and B', and if a, b, and c represent the concentrations of these three constituents in a given solution, then the following simple formula holds good—

$$ab = kc.$$

Now the two kinds of ions are produced in equivalent quantities in the above case, hence $a = b$. If, further, the total amount of the electrolyte $= 1$, and a represents the ionised portion, then $a = b = \dfrac{a}{v}$ and $c = \dfrac{1-a}{v}$, v being the volume of the solution in which unit quantity (a molecular weight in grammes) of the electrolyte is contained. By carrying out the substitution we get the formula

$$\frac{a^2}{(1-a)} = kv,$$

which expresses the state of ionisation of an electrolyte at the dilution v.

We gather from this formula that a must become larger as the dilution v increases; when that dilution becomes infinitely great, then $1-a=0$, or $a=1$, *i.e.* the electrolyte is completely ionised. On the other hand, the value of a becomes very small when v approximates to nothing, *i.e.* at maximum concentration there is minimum of ionisation. For the rest, the state of ionisation at a given dilution v depends upon the value of the constant k. This is very high and much the same for all neutral salts, but it varies enormously in the case of acids, being high for strong acids and low for weak.

The differences in the degree of ionisation of the various acids become less and less marked with increasing dilution of their solutions, whence it follows that infinitely dilute solutions of different acids possess the same strength, because the acid in them is completely ionised. And the same applies to bases, which also show very considerable differences in the constant k.

8. *Complex Dissociation*

The ions of salts are by no means to be regarded as absolutely stable compounds; on the contrary, they may readily and in various ways undergo hydrolysis, or ordinary or electrolytic dissociation. Thus the metallo-ammonium ions,[1] for example, usually dissociate more or less into metallic ion and free ammonia, and the complex ion $Ag(CN)_2$ of potassium-silver cyanide is dissociated electrolytically into $Ag^{\cdot}$ and $2CN'$, etc. etc. The laws governing those dissociations are precisely the same as hold good for the forms of

[1] *e.g.* $Ag4NH_3$, etc.

chemical equilibrium which have been already discussed, and therefore require no further explanation. We must, however, be careful to bear well in mind the possibility of such changes, if we wish to thoroughly understand the ofttimes complicated play of chemical equilibrium.

9. *Gradual Dissociation*

In the dissociation of electrolytes, which are made up of ions of unequal valency, *e.g.* of the dibasic acid, H_2A, one would be inclined to assume the validity of the following reaction—

$$H_2A \rightleftarrows 2H^{\cdot} + A''\!,$$

whence would follow an equation of the form—

$$ab^2 = kc.$$

But experience shows that this is not the case. The dibasic acids really dissociate according to the formula—

$$H_2A \rightleftarrows H^{\cdot} + HA'\!,$$

and the resulting monovalent ion itself experiences the further dissociation—

$$HA' \rightleftarrows H^{\cdot} + A''\,;$$

the dissociation-constant of this second reaction being in every case very much smaller than that of the first.

It follows from this that the various hydrogen atoms of a polybasic acid have different values as regards the "strength" of the latter; the first atom of hydrogen will always be that of a stronger acid than the second, while the third and fourth, etc., will follow in the same order.

10. *Several Electrolytes together*

The same equilibrium-formula suffices for the reciprocal action of several electrolytes present in the same solution together; a few important cases of this will be instanced here.

Two neutral salts exert almost no action upon one another, because both they themselves and the possibly resultant new salts produced by their double decomposition are all strongly ionised, and the ions remain substantially in their original condition. Thus, a solution of potassic chloride consists chiefly of the ions $K^{\cdot}$ and Cl', and one of sodic nitrate of the ions $Na^{\cdot}$ and NO_3', and these conditions are not altered by mixing the one solution with the other. Further, this resulting solution is necessarily identical with one made from corresponding quantities of sodic chloride and potassic nitrate, seeing that the former contains the same ions in the same free state as the latter.

An action takes place, however, if the ions present are capable of uniting to form one or more compounds, which are either not dissociated at all (practically speaking) or only slightly so under the existing conditions. The constant k has then a small value; in the equation

$$ab = kc$$

a and b (the concentrations of the ions) must therefore become greatly reduced, while c (the concentration of the non-dissociated portion) must grow correspondingly until the equation is fulfilled. By bringing in the dissociation factor a, we then get

$$\frac{a^2}{1-a} = kv \; ;$$

the value of v being constant, and k and a being both low, the amount of electrolytic dissociation must be small.

The reaction thus results in the more or less complete disappearance of the ions of the electrolyte, which has a small constant k, going as they do to build up the non-ionised portion.

The most characteristic case of such an interaction is the neutralisation of an acid by a base. An acid contains an anion and hydrogen, $H^{\bullet}$, while in a base we have a cathion and hydroxyl, OH'; water, the compound of hydrogen and hydroxyl, is ionised to but a very slight extent, and must therefore be formed as soon as its two constituent ions meet in any liquid. Hence it follows that when solutions of acids and bases are mixed together we get an energetic interaction, the hydrogen and hydroxyl ions uniting to form water, while the two other ions of the resulting salt remain in the solution in the free state.

We get the same kind of phenomenon upon adding a strong acid to the salt of a weak one. As has been mentioned already, neutral salts are all ionised to about the same extent, however strong or weak their acid may be. The solution of a salt of a weak acid therefore contains for the most part free ions only; when to this solution there is added a strong acid, which is likewise almost completely ionised, the anions of the salt unite with the hydrogen of the acid to form a second acid which is but little ionised, seeing that we have started with the salt of a weak acid, *i.e.* of one which undergoes only slight ionisation. At the same time the anion of the added (strong) acid remains along with the cathion of the original salt; in other words, the solution now contains the (ions of the) salt formed by the production of the weak by means of the strong acid. The real engendering cause of the reaction is thus not what has hitherto been supposed, *i.e.* it does not lie in the "attraction" of the stronger acid for the metal, but in the tendency of the ions of the weak acid to pass into the non-ionisable condition.

The above reaction is not carried out with such completeness as the formation of water on neutralisation, because even weak acids are always more ionised than water; the "expulsion" is always less perfect the more strongly the newly formed acid is ionised. Should the latter be ionisable to the same extent as the added acid, nothing of course will take place.

Precisely the same considerations hold good for the action of a strong base upon the salt of a weak one. What occurs when one or other of the substances is sparingly soluble will be explained later on.

11. *Acids and their own Salts*

Considerable action goes on in a case where it was not formerly suspected, viz. on mixing solutions of salts and acids containing the same anions, or—to put it generally—when two electrolytes with a common ion meet in solution.

If both electrolytes are equally dissociated there is of course no action of any consequence, but the case is different when a slightly dissociated electrolyte, *e.g.* a weak acid, meets with a strongly dissociated one containing the same ion, *i.e.* with a salt of the acid in question. The result is that there is always a more or less marked retrogression in the dissociation of the weak electrolyte. From this the rule follows that moderately strong or weak acids act much more feebly in presence of their own neutral salts than they do in the pure state, equal concentration and acidity being of course taken for granted.

To understand this we have only to remember that the state of equilibrium of the partially ionised acid is given by the equation

$$ab = kc,$$

in which a is the concentration of the anion, b that of the cathion (hydrogen in this case), and c that of the non-ionised portion; c being large in comparison with a and b in the case of weak acids. If a neutral salt of the same acid—containing the same anion—be now added, a will become greatly increased, while b must be lessened in almost the same ratio, since c can be only slightly augmented, from the fact of the greater part of the acid being already present in the non-ionised state. The hydrogen ions therefore diminish materially. Now it is precisely upon the concentration of the hydrogen ions that the characteristic reactions of acids depend. The more of the neutral salt therefore that is added to them, the more are those reactions weakened; and (what follows likewise from the above considerations) the effect is the more marked the weaker the acid.

These conditions are of very frequent occurrence in analysis, especially in those cases where an acid reaction but the least possible acid action is desired. In such cases (*e.g.* in the precipitation of zinc by sulphuretted hydrogen), if the solution contains a strong acid like hydrochloric, it is usual to add to it an excess of sodium acetate. In this way we not merely substitute the slightly ionised acetic for the strongly ionised hydrochloric acid, but there is the further effect—that

F

the ionisation of the acetic acid itself is lowered to a very considerable extent. Such an addition as this has also the result of yielding a liquid which behaves almost as if it were neutral, although it has an acid reaction; and it retains this approximate neutrality even although a free strong acid continues to be liberated, by the action of the sulphuretted hydrogen in the example just cited. For this (strong) acid undergoes instant transformation, as shown above, and thus the concentration of the few hydrogen ions is only increased in relatively very slight degree.

Considerations of the same kind present themselves when a weak base and one of its neutral salts are present together in the same solution. The reciprocal action between a strong acid (or base) and a weak one likewise falls under the same category, the ionisation of the weaker constituent being always lowered. Such cases occur less frequently, however, in actual analysis.

12. *Hydrolysis*

Although water is extremely undissociable, recent investigations have proved that it is in fact broken up to a certain definite extent into hydrogen and hydroxyl ions, and the amount of this dissociation or ionisation has been measured. It has thus been shown that water contains a gramme-equivalent of its ions in a million litres approximately.

In consequence of this the process of neutralisation, described on p. 63, does not complete itself absolutely, but as many hydrogen and hydroxyl ions remain uncombined as are normally present in water. This residue—as already seen—is extremely small, and may be neglected in most cases. Still there are conditions possible under which this trifling quantity exercises a measurable effect, and those occur when the acid or the base or both of these are very slightly ionisable or very weak.

According to the laws of chemical equilibrium, therefore, the effect of the presence of hydrogen ions in the solution of a neutral salt is that not merely the free anions of the salt, but also a corresponding quantity of non-ionised acid molecules are present, in accordance with the formula $ab = kc$. If now k has a high value, as in the case of strong acids, c is very small, seeing that b—the concentration of the hydrogen ions —is small. If on the contrary k has a low value, c— the concentration of the non-ionised portion of the acid—increases correspondingly, and approximates to k in its numerical value in the ionisation constant of water; c thus becomes measurable, *i.e.* the presence of non-ionised acid in a solution of its neutral salt becomes possible. Cyanide of potassium may be taken as an instance of this; hydrocyanic acid has an extremely small ionisation-constant, and an aqueous solution of potassium cyanide therefore contains a measurable quantity of the non-ionised acid, as we can readily perceive from the odour.

Another thing which is characteristic of such salts is their alkaline reaction. The latter depends upon the presence of hydroxyl ions, and, in order that it may be noticeable, the concentration of these ions must reach beyond a given point, which varies with the sensitiveness of the indicator used. Now we have just seen that, in the case of salts of weak acids, a certain quantity of non-ionised acid results, the requisite hydrogen for which is taken from the water. And since in water, which is a substance of constant concentration, the product of the concentrations of the hydroxyl and hydrogen ions must have a constant value (according to the law of equilibrium), if the concentration of the latter is diminished to the nth part, that of the former must grow n-times, and thus become measurable when n is a large number.

Exactly the same considerations apply to salts of weak bases. These will have an acid reaction, and the presence of non-ionised base will be recognisable.

Should both acid and base be weak, the processes which have just been explained will mutually support one another, in the sense that appreciable quantities of non-ionised acid and base will result; the production of an excess of hydroxyl and hydrogen ions will on the other hand be diminished, since the cathions of the base will use up the one, and the anions of the acid the other.

13. *Heterogeneous Equilibrium; Law of Distribution*

If the structure in which equilibrium obtains is divided into several portions by surfaces of separation, the law holds that—*in two contiguous spaces or phases, the concentrations of each substance, which is present in both spaces, must bear a constant ratio.* If a' be the concentration of a substance A in the first space, and a'' the concentration in the second, then

$$a' = k a'',$$

k being a coefficient depending upon the nature of the substance and the temperature.

Similar equations are to be set up for each constituent present. Here, again, we have to bear in mind that ions are to be treated as separate entities; and, further, different modifications of the same substance are to be regarded as different substances.

What has been said with respect to the preceding law applies also to this one; it is a law limited to dilute solutions or gases, the concentration function for strong solutions being unknown.

Particular cases of this law have already been dealt with. Thus the law of absorption of gases (p. 39) is one of these, as can be seen on comparing the two formulæ. The law of the solubility of solids in liquids and the law of vapour pressure also belong to the same category. In both of these cases the state of the body in one of the two phases remains the same throughout; the solid in contact with its solution and the liquid in contact with its vapour alter indeed in amount but not in condition, and therefore what we term their concentrations remain unchanged also. One of the two factors a' or a'' thus continues constant in the equation, and the other must necessarily remain constant also; hence we have for each substance a certain definite solubility and definite vapour pressure, depending upon the nature of the substance and the temperature, but not upon the amount of fluid or solid and the volume of vapour dealt with.

The same state of things also frequently occurs in the case of homogeneous equilibrium. It is therefore convenient to draw a distinction between *states of constant concentration and states of variable concentration*. Solids possess constant concentrations, and also simple liquids (not mixtures of liquids); gases and dissolved substances show variable concentrations. We may also rank among substances of approximately constant concentration such constituents of liquid or gaseous mixtures as are present in very large amount compared with the others; true, in the course of a reaction their concentration alters, but this alteration is the less the more they preponderate over the other constituents, and they may thus be looked upon in many cases as being practically pure substances.

These two simple laws of mass-action and of distribution embrace in principle all the phenomena of chemical (including physical) equilibrium. There will be frequent opportunity in what follows of this book to elucidate the abstract theory by investigating particular cases.

§ 3. THE COURSE OF CHEMICAL REACTIONS

14. *The Velocity of Reaction*

In addition to having a knowledge of the laws of chemical equilibrium, it is necessary that the analyst should be able to follow the progress of chemical reactions as well. For, although most of the analytical methods in vogue are ion-reactions, which complete themselves in an immeasurably short space of time, there are certain processes which do not come into this category, and which we cannot criticise without having the above information.

A similar expression to that given for equilibrium holds good for the velocity of a reaction, since the state of equilibrium may be defined as that state in which the velocities of opposing reactions are equal. In other words, the velocity of a reaction is directly proportional to the concentration of each reacting constituent; if several molecules of a substance are taking part, the concentration of the latter is to be raised to the corresponding power. By the velocity of reaction is to be understood the ratio between the transformed substances and the time required for transforming them. The amounts of the substances are to be calculated here, as always, in units of molecular and not of absolute weight.

The different cases of course of reaction, arranged according to the number of reacting substances, and according to the original ratio of their amounts, have this in common—that they begin with their greatest velocity-value, and that the velocity becomes less and less as the reaction proceeds. They all lead to the theoretical result—that the reaction is only complete after an infinite length of time. But for practical purposes we may apply the rule that any residual action may be regarded as non-measurable, after ten times the period which is required to complete the first half of the reaction.

15. *Influence of Temperature*

Temperature has an extraordinarily marked effect upon the velocity of chemical reactions; in the instances where measurements have been made the velocity has doubled itself with a rise of a few degrees of temperature. In every case of slow chemical reaction the end will be reached far more quickly with the aid of heat.

16. *Catalysis*

There are cases in which certain substances exercise quite an exceptional influence upon velocity of reaction, although they appear to take no part in the process itself; such actions are termed catalytic, and the substances in question catalysers. Besides specific catalysers, effective in certain definite reactions (*e.g.* the salts of iron in oxidations and reductions), the acids may be looked upon as general catalysers. We may state it as a law that slow reactions are invariably hastened by the presence of acids (assuming that the latter form no chemical compounds with the reacting substances), and this effect is proportional to the "strength" of the acids, or—to speak more correctly—proportional to the concentration of the free hydrogen ions in the liquid. For example, the conversion of pyro- or meta-phosphoric acid in aqueous solution into the ortho-variety is immensely facilitated by the presence of nitric or hydrochloric acid, while the slightly ionisable acetic acid is almost without effect.

Catalytic actions are also known among gases; here, however, they are often induced by chemically indifferent substances possessing a large superficies, and are thus more of a mechanical nature. Finely divided platinum or palladium has a wonderful effect in quickening processes of combustion.

17. *Heterogeneous Structures*

The foregoing remarks apply to homogeneous structures. In heterogeneous the velocity of reaction is further dependent upon the extent of the contact surface. Since the reaction only takes place at the contact surface itself, where the velocity rapidly diminishes on account of saturation, a thorough mechanical mixing of the whole reacting mass is necessary to hasten the process, care being of course taken to ensure the greatest possible extent of surface at the commencement, either by powdering or other suitable means.

§ 4. PRECIPITATION

18. *General Considerations*

It has been already mentioned at a previous stage that, of the possible combinations of states of aggregation for separating substances from one another, the solid-liquid is the easiest to manage and also the most perfect for practical purposes. Our preparatory chemical measures are therefore mainly directed to bringing this combination about, and *precipitation* is one of the most common procedures in chemical analysis.

Precipitation follows when the constituents of some substance, which is not completely soluble under the circumstances, come together in solution. Each precipitation is thus preceded by a state of *super-saturation*, and after it is complete the liquid is saturated with respect to the precipitated solid, *i.e.* it is in equilibrium with the latter. As a matter of fact no precipitation is ever perfect, and it is therefore the aim of the analyst to reduce the amount of dissolved residue to the minimum.

19. *Super-saturation*

When a solution contains more of a solid or of its constituents than accords with the state of equilibrium, it is said to be super-saturated as regards that solid. The deposition of the latter from such a solution need not follow so long as there is no trace of the substance present in the solid form, and we can thus often keep a super-saturated solution for an indefinite period, if care be taken to prevent any such contact. Should, however, some of the solid be present, precipitation must necessarily go on until equilibrium is established. But, as this ensues only on the contact surfaces between the solid and the solution, it is possible even under those circumstances to retain the state of super-saturation for a long time, provided that the contact surface is small and that mechanical movement is avoided.

On the other hand, deposition may come about in a super-saturated solution without any of the solid being present. And this occurs the more easily and surely the greater is the ratio between the momentary concentration and the final concentration corresponding to equilibrium. Deposition is further often promoted by vigorous shaking, stirring, etc.

It follows that—under otherwise similar conditions—super-saturation is easier to bring about, the more soluble the solid is. The three sulphates of barium, strontium and calcium furnish a good instance of this; while the precipitation of the first of these salts is almost instantaneous, even from very dilute solutions, that of the second requires a measurable time, and in the case of sulphate of calcium a moderate super-saturation may last for weeks and months. It is, however, obvious that the special nature of the substance exerts an influence here, thus causing certain compounds to show the phenomena of super-saturation very readily, while with others it is very hard to get them.

The most effective way of ending a state of super-saturation is to shake the solution thoroughly with some fragments of the solid in question, or to continue stirring for a long time after a precipitate has once formed. Speaking generally, this putting an end to super-saturation is a time-phenomenon of the general character described on pp. 70-72.

20. *The Solubility-Product*

It is only in very rare cases that precipitates obtained in analysis can be redissolved in the unaltered state. On the contrary, they are almost without exception electrolytic in their nature, and their aqueous solutions contain mainly the ions of the compound, together with a very small portion of non-ionised salt. Since we are dealing here with very difficultly soluble substances, we may always regard their solutions as—for practical purposes—completely ionised.

In order therefore to ensure the fullest practicable separation of his precipitate, the analyst has to establish such a condition in the solution that the latter shall dissolve the precipitate to the least possible extent. In the case of unaltered soluble or indifferent substances (*e.g.* sugar), this is achieved on the one hand by lowering the temperature, and on the other by the addition to the solution of something which serves to diminish its solvent action, *i.e.* of some substance in which the solid is even less soluble than in the main constituent of the liquid. We can in this way, for example, separate many organic compounds from solution in ether by adding petroleum, or resin from solution in alcohol by adding water.

When the precipitate is an electrolyte, we have a very perfect means at command for diminishing its solubility, viz. *by the addition of another electrolyte, which has an ion in common with the precipitate.*

In the saturated aqueous solution of an electrolyte we have a complex equilibrium. On the one hand the solid is in equilibrium with the non-ionised portion of itself which is in solution, while on the other this non-ionised portion is in equilibrium with the dissociated part, *i.e.* with the ions of the same substance. The first equilibrium comes under the law of proportional concentration, or, since we are dealing here with a substance of unalterable concentration on the one hand, the concentration of the non-ionised portion in the solution must have a perfectly definite value. For the second equilibrium we have in the simplest case, *i.e.* when the ions of the compound are monovalent—

$$ab = kc,$$

a and b representing the concentrations of the ions and c the concentration of the non-ionised portion (see p. 59).

Now since c is constant at a given temperature, as we have already seen, kc and therefore ab must be constant also. Equilibrium is thus established between a precipitate and the liquid above it when the product of the concentrations of the two ions, into which the precipitate falls, has a definite value. This product may be termed the *solubility-product* for the sake of brevity.

If the electrolyte consists of polyvalent ions in the proportion $mA : nB$, the solubility-product takes the form :

$$a^m b^n = \text{a constant.}$$

Whenever in any liquid the solubility-product of a solid is exceeded, the liquid is super-saturated with respect to that solid; but whenever the solubility-product is not yet reached, the liquid exerts a solvent action upon the solid. The whole theory of precipitates is comprised in these simple laws, and all the accompanying phenomena—the diminution as well as the so-called abnormal increase in solubility—can not merely be explained by their aid, but can be predicted when the conditions are given.

In considering the application of the law with respect to the completeness of the precipitation of any given substance, we have to remember that the analyst's task is always to separate a definite ion. Thus a precipitate of barium sulphate is produced with the object of determining either the sulphuric acid ion, SO_4'', or the barium ion, $Ba^{\cdot\cdot}$, precipitation being brought about in the first case by adding a barium salt, and in the second by adding a sulphate. Let us consider the former. If we add exactly the equivalent in barium salt of the SO_4'' present, some of the latter remains in solution—as much, namely, as will yield the solubility-product of barium sulphate with the $Ba^{\cdot\cdot}$ ions which are also present. But if more barium salt be now added, the corresponding factor of the product will be increased; the other must therefore become smaller, and some more sulphate will be thrown down. Further addition of barium salt increases the effect still more, but the quantity of SO_4'' ions can never be reduced to zero, since it is impossible to make the concentration of the $Ba^{\cdot\cdot}$ ions infinitely great.

This puts us in a position to appreciate the old-established rule, always to use an excess of precipitant. But a further rule follows from the above considerations also, viz. that this excess must always be greater, the more soluble the precipitate is. For, in order to reduce the concentration of the precipitable ion to the nth part of that which it has in the pure aqueous solution of the precipitate, n-times the quantity of the other ion are required; the amount of the latter must thus be increased in that proportion. If, therefore, we set ourselves to reduce the solubility down to a given *absolute* amount of the precipitable ions, the concentration of the precipitating ions must be multiplied by the ratio of the two solubility-products, in order that the end may be attained.

In the case of most precipitates a moderate excess of precipitant is sufficient for the purpose. Of course, if a precipitate is to be of any use for analytical purposes, it must have a small solubility-product.

What holds good for the throwing down of a precipitate also retains its significance for the washing of the latter. If the precipitate should be appreciably soluble in pure water, loss can be avoided by washing with a solution containing an ion in common with the precipitate. Thus sulphate of lead is better washed with dilute sulphuric acid than with pure water, and mercurous chromate with a solution of the nitrate. For obvious reasons these wash-liquids are most conveniently made of a dilute solution of the precipitant; they are so chosen as to cause the least possible inconvenience in the subsequent treatment of the precipitate.

21. *Some Precipitation-Reactions*

Precipitation always follows in a liquid, in accordance with the above laws, when ions which belong to a substance of less solubility or of small solubility-product meet in a liquid. The relations are simplest in the case of neutral salts, which—as already stated on p. 49—are strongly ionised to about the same extent more or less. It is thus sufficient to bring together two salts, each of which contains one of the ions in question alongside of any other; *e.g.* a neutral barium salt yields a precipitate of barium sulphate with any neutral sulphate.

The relations are more complex when acids or bases come into play, since in their case we may have any degree of ionisation, from the strongest to the weakest, whereby the conditions may be such that there is no precipitation at all, although this takes place when solutions of the corresponding neutral salts are employed. Thus calcium salts are thrown down by all carbonates, while free carbonic acid has no effect upon them. This arises from the soluble carbonate being ionised normally; if a solution of the latter is added to a solution of a calcium salt, the product of the concentrations of the carbonic acid and calcium ions is far greater than the solubility-product of calcic carbonate, and precipitation ensues. On the other hand, carbonic being an excessively weak acid, its aqueous solution contains only a minute proportion of carbonic acid ions; and thus, notwithstanding the abundance of calcium ions, the critical value of the solubility-product is never reached, and consequently no precipitate of calcium carbonate can form.

The case of lead salts is somewhat more complicated. Carbonate of lead is less soluble than carbonate of calcium, and therefore when carbon dioxide is led into a moderately concentrated solution of a lead salt, the value of the solubility-product is reached, in spite of the small ionisation of carbonic acid into CO_3'' and $2H^{\bullet}$; hence precipitation ensues. This causes the disappearance of $Pb^{\bullet\bullet}$ ions on the one hand and of CO_3'' ions on the other, while there remain the hydrogen ions from the ionised carbonic acid and the anions of the lead salt—$2NO_3'$ if lead nitrate is used—*i.e.* free nitric acid is produced. If the reaction is carried further, the concentration of the latter, *i.e.* of the hydrogen ions, increases; these hinder the ionisation of the carbonic acid (cf. p. 62), so that a point is ultimately reached at which no more carbonic acid ions can be formed, and consequently no more carbonate of lead precipitated.

It depends upon the anion of the lead salt to what extent the reaction goes on. If it is one of a strong acid, the hydrogen ions remain in that state, and soon attain to the critical concentration. If, on the contrary, it is one of a weak acid, then the hydrogen ions unite in greater or less degree with those anions to non-ionisable acid, and the decomposition goes much further. Thus acetate of lead is precipitated by carbonic acid to the extent of two-thirds, while the nitrate is only just affected.

If some strong acid be added beforehand to the solution of lead salt, precipitation will be altogether prevented, since the presence of the hydrogen ions will stop the carbonic acid from dissociating far enough to yield the value of the solubility-product with the lead ions. On the other hand, the decomposition of a solution of lead acetate can be materially increased by adding another soluble acetate to it. This increases the acetic acid anions, which are thus enabled to bind far more of the liberated hydrogen anions to non-ionised acetic acid, before these latter reach their critical concentration and thus prevent the ionisation of the carbonic acid, and with this the precipitation of carbonate of lead.

Precisely similar conditions apply to the precipitation of metallic salts by sulphuretted hydrogen, which is such an important procedure in analysis. As this will be discussed in detail in the special part of the book, no further reference need be made to it here.

The conditions which regulate the differences in the action of basic precipitants are likewise the same. The strongly ionised caustic potash precipitates all the sparingly soluble hydroxides, while the slightly ionised ammonia can only bring down the weakly basic among them; the latter, for example, cannot precipitate the hydroxide—sparingly soluble though it is—from calcium salts.

22. The Redissolving of Precipitates

The laws governing the solubility-product also furnish us with a complete answer to the query—What are the conditions under which precipitates become soluble again? We may expect that every cause which diminishes or practically eliminates one of the substances from the solution (i.e. one of the ions or the non-ionised portion) must increase the solubility of the precipitate. And it is the fact that when a substance producing this effect is added, so much of the precipitate goes into solution as allows of the product regaining its original definite value.

The simplest and most common case in point is the solution of an "insoluble" base in an acid. When, for example, magnesium oxide is rubbed up with water, it forms a solution containing the ions of magnesium and hydroxyl, together with a very small quantity of non-ionised hydroxide. If an acid like hydrochloric is now added, whose solution consists substantially of hydrogen and chlorine ions, the hydrogen and hydroxyl ions immediately join together to form water.[1] The product of magnesium and hydroxyl[2] therefore becomes too small, and more magnesia passes into solution, the process repeating itself continuously. This goes on until all the hydrogen ions of the hydrochloric acid are used up; in the solution we find the corresponding quantity of magnesium ions together with the unaltered chlorine ions, *i.e.* magnesium chloride. Less magnesium hydroxide naturally dissolves in the solution of this salt than in pure water, since there is now a large excess of magnesium ions present.

The action of a soluble base upon a sparingly soluble salt, with which it unites to form one easily soluble, is explicable in exactly the same way.

The action, too, of acids upon many difficultly soluble neutral salts is dependent upon the same causes. When, for instance, hydrochloric acid acts upon calcic phosphate, the ion of the phosphoric acid present in solution unites with the hydrogen of the hydrochloric acid to form non-ionised phosphoric acid for the most part, the latter being much less ionisable than the halogen acid. Phosphoric acid ions thus disappear, and new calcium phosphate must go into solution, and so on. This case differs, how-

---

[1] As has been mentioned already on p. 66, water is only ionisable to an exceptionally slight extent, and it is therefore always produced when hydrogen and hydroxyl ions meet.

[2] Strictly speaking, the product of the magnesium and the square of the hydroxyl, the formula being—

$$Mg(OH)_2 \rightleftarrows Mg^{\cdot\cdot} + 2OH'.$$

ever, from the preceding one in that the hydrochloric acid cannot dissolve quite its equivalent of calcium phosphate. For, since the phosphoric acid ionises of itself, albeit to a far less extent than the hydrochloric, its anions are not so completely used up as those of hydroxyl were in the former case, but accumulate the more in the solution, the more hydrogen ions of the hydrochloric acid have been already consumed. It is ultimately present in such large quantity as to give with the augmented calcium ions the critical value of the product, when the solvent action of the hydrochloric acid ceases, although free hydrogen ions are still there.

It is thus obvious that an essential condition of the reaction consists in the resulting acid being but slightly ionisable. In other words, only sparingly soluble salts of *weak* acids, and not salts of strong ones, are dissolved by strong acids. This deduction is completely confirmed by experience; the halogen compounds of silver, the sulphates of barium and lead and other compounds of strong acids are insoluble in dilute acids, even when the latter are among the strongest known and form soluble salts with the metals of the precipitates in question. On the other hand, all salts of the weaker acids are soluble in strong acids, and this the more, *ceteris paribus*, the weaker the acid is. Thus, most of the phosphates dissolve readily in acetic acid, while the oxalates—as salts of a stronger acid—dissolve but sparingly, although readily in hydrochloric acid. Since, however, the solubility of precipitates depends not merely upon this, but also upon the numerical value of the solubility-product, the subject is more complex than one would expect from what has just been said.

Exactly the same considerations apply to the solution of insoluble salts of weak soluble bases in strong bases; such instances are, however, not of common occurrence.

The cases which have been discussed above are not the only ones in which insoluble precipitates are brought into solution by reagents, for the ions of these may experience other fates than transformation into water or into non-ionised acids or bases. Every reaction which tends to diminish the number of free ions produces a like effect. A few phenomena in point will therefore be taken now in order that we may get a general idea of the subject; all the more important analytical reactions of this kind will be considered in detail in the special part of the book.

Alumina dissolves readily in alkalies, but is very slightly soluble in water. The aqueous solution contains the ions $Al^{\cdot\cdot\cdot}$ and $3\,(OH)'$, and the precipitate is in equilibrium with these as well as with the non-ionised dissolved aluminium hydroxide. On the addition of potash the latter changes into potassic aluminate, AlO_3K_3, whose ions are AlO_3''' and $3K^{\cdot}$, and a part of the suspended alumina must dissolve (in water) in order that the equilibrium may be re-established. This aqueous solution is in its turn attacked by potash, and the interactions go on repeating themselves until the potash is no longer capable of transforming more aluminium hydroxide into aluminate. The action in this case thus depends upon the conversion of the cathion $Al^{\cdot\cdot\cdot}$ into the anion AlO_3''', and the loss of the former as regards the equilibrium.

The solvent action of ammonia upon the salts of copper, silver, nickel, etc., is still simpler. In the first place the hydroxide in question is precipitated (as given on p. 81). On further addition of precipitant the metallic ion unites with the excess of ammonia to form a compound ion of the general formula $M(NH_3)_n$; the equilibrium is disturbed by the disappearance of the metallic ion, hence new hydroxide goes into solution, the metallic ion of which is in its turn used up, and so on.

The solubility of many otherwise insoluble metallic compounds in cyanide of potassium can be explained in the same way. Thus ferrous hydroxide acts upon potassic cyanide to form the ferrocyanide and free potash—

$$Fe(OH)_2 + 6KCN = K_4Fe(CN)_6 + 2KOH.$$

The reaction is that the ferrous ion unites with the 6CN to the anion of ferrocyanide of potassium, so that as the ferrous hydroxide passes into solution it is immediately taken up, and more of it must go on dissolving until the amount corresponding to the formula has been converted. The solution contains no ferrous ion demonstrable by analysis, for it gives none of the reactions which are characteristic of ferrous salts.

Further cases will be considered in the special part of the book.

§ 5. REACTIONS ATTENDED WITH THE LIBERATION OR ABSORPTION OF GAS

23. *The Liberation of Gas*

The second case of separation, *i.e.* the separation of liquids from gases, is much less frequent in analysis than precipitation. Two opposite procedures have to be distinguished here, viz. either liquids alone are present, and one of them has to be converted into gas, or we have a mixture of gases, one of which has to be transformed into liquid or solid.

For the liberation of a gas from a liquid, in which the former is present either actually or potentially, the laws of heterogeneous equilibrium apply just as in precipitation. Here, however, we have not got the simplifying factor that came into the latter calculation, viz. that one of the substances possesses a constant concentration. True, so long as a gas is pure and under a constant pressure (*e.g.* that of the atmosphere), it may be looked upon as being of constant concentration. But when another gas is admixed with it, the concentration or the partial pressure can easily be reduced to any value, and in this possibility there lies an important aid to analysis.

The phenomena of super-saturation can be brought about very readily in solutions of gases, as in those of solids; if the super-saturation is only slight, it may last a long time, but if considerable, it ends itself spontaneously, when we have effervescence taking place. The mode of putting an end to it is to bring the solution into contact with *any* gas, being thus of more general application than in the case of solutions of solids. When the super-saturated solution is thus in contact with some gas, the gas which is dissolved diffuses into this, and the process is the more rapid the greater the degree of super-saturation. Thus every bubble of gas in the liquid aids in the separation of further gas. Practical use is made of this, in order to rid a liquid of the last portions of a gas which does not escape spontaneously, by passing a current of some other indifferent gas through the solution.

The amount of gas which without this means of expulsion would otherwise remain in the solution is proportional to the pressure, the absorption-coefficient of the gas in the liquid (which diminishes as the temperature rises), and the volume of the latter. A spontaneous escape of gas or effervescence will thus only take place if the amount of gas in question is appreciably greater than what would be dissolved under the above conditions. It is therefore advisable, in those cases where only a little gas is to be looked for, to work with as concentrated solutions as possible and at a raised temperature.

Gases which dissociate for the most part into ions, when dissolved in water, cannot be expelled from moderately dilute solutions. The halogen hydracids are examples of this. In order to obtain such substances in the form of gas, they must be generated under conditions which make the dissociation into ions impossible, or at least reduce this to its lowest limit; and water, more especially, must be absent. The expulsion of hydrochloric acid gas from its aqueous solution by the addition of concentrated sulphuric acid (a method which one sometimes uses for purifying crude muriatic acid) depends upon the reproduction of non-ionised hydrogen chloride, as set forth on p. 65.

In accordance with this, all those gases which can be completely expelled from their aqueous solutions are either indifferent in their nature, or, should they be acid or basic, they yield only weak acids and bases. Ammonia and sulphur dioxide stand approximately at the outside limits of these. This gives us a guide with respect to the conversion of any given substances into gases for purposes of separation—the resulting gas must be as far as possible of an indifferent nature, for ions are not volatile. A slight ionisation makes this separation more difficult but not impossible. For, although only the non-ionised part assumes the gaseous form and can thus be expelled, the diminution of this portion alters the ionisation-equilibrium continuously, in that fresh non-ionised substance is built up at the cost of the ions, until these finally disappear.

24. *The Absorption of Gas*

The reverse applies to the chemical absorption of a gas. We have to get it into the ionic state; hence acid gases must be absorbed by alkaline liquids and basic gases by acid ones. To bring about the absorption of an indifferent gas from a mixture is much harder, since all reactions in which non-electrolytes take part proceed more slowly than ion reactions. For the rest, the most thorough possible contact must be brought about, by one or other of the means which have been already detailed.

The chemical absorption of a gas by a solid takes place under like conditions. The essential *rôle* of ions becomes apparent here from the fact that with absolutely dry substances the usual reactions between gases alone and between gases and solids generally remain in abeyance. It is not, however, necessary as a rule to observe any particular precautions in this respect in an analysis, since most substances attract sufficient moisture from the air during the operation to allow of the production of the very small quantity of ions necessary to the commencement of the reaction. But whether this explanation holds good in every case of the kind requires further investigation.

§ 6. REACTIONS ACCOMPANYING THE EXTRACTION OF A DISSOLVED SUBSTANCE FROM ONE SOLVENT BY MEANS OF ANOTHER.

25. *Influence of the Ionic State*

In separations which are brought about by the aid of two non-miscible liquids, we have always an aqueous solution to deal with on the one hand, and it is therefore well to bear in mind that ions will no more leave the aqueous solution in such a case than they will assume the gaseous form. In order, therefore, to separate a substance from its solution in water by ether, benzene, or any such liquid, it must be brought into a condition in which it is neither an ion itself nor a constituent of one.

What was said on p. 87 applies to partially ionisable substances also. It is only the non-ionised portion which is affected by the process of extraction, and therefore the distribution-coefficient and the law given on p. 41 apply to it only. So, to "shake up" solutions of such substances to the best advantage, the conditions must be regulated with the view of having the non-ionised constituent present to the greatest possible extent. The aqueous solution must therefore be made as concentrated as circumstances will permit, and it is a great additional help, when dealing with acids of only moderate strength, to add some very strong acid such as hydrochloric, and when dealing with bases of moderate strength, some alkali. It follows from what was said on pp. 64-65, that such additions will increase the amount of non-ionised substance, and hence a given quantity of the second solvent will withdraw from the aqueous solution more of the substance in question than it could otherwise do.

We may look forward to the methods of separation by extraction by means of a solvent—*e.g.* with regard to the alkaloids—becoming further elaborated than they are at present. More particularly we may expect distinctions to be drawn between a strongly acid solution (hydrochloric acid) and one weakly acid (acetic acid plus sodium acetate), of which the one will retain the substances that can be extracted from the other.

§ 7. THE ELECTROLYTIC METHOD

Electrolytic methods differ from all the other methods of separation which have been already discussed, in that the chemical transformation and mechanical separation are in their case comprised in one act. The procedure is based upon the fact that, by the action of an electric current, the positively electrified cathions move in the direction of the positive slope of the electric potential, and the anions in the opposite direction. So long as those movements take place inside the electrolytic liquid, they are governed by the law that equal quantities of positive and negative electricity must be present together in any given space, and therefore also equivalent quantities of positive and negative ions. For, according to Faraday's law, equivalent quantities of ions—be they what they may—are combined with like amounts of electricity, the cathions with positive and the anions with negative. So long therefore as the current passes only inside the liquid, it effects no separation; the ions do indeed displace one another, but each space in the liquid gets filled up with exactly the same number of ions of either kind, as have been driven out of it by the action of the current.

The conditions are, however, entirely altered when the electric current is caused to flow to the outside of the liquid. Here again equal amounts of positive and negative electricity must of necessity leave the liquid at the same moment, the point at which the former takes place being named the cathode and the other the anode. And, since electricity leaves the liquid, the corresponding quantity of ions which were charged by it must be transformed into the non-electric or non-ionic state. This takes place only at that spot where the electricity leaves the electrolyte, and hence electrolytic reactions go on solely at the surface of contact between the electrodes and the electrolyte.

The reaction which takes place at the electrodes is not always of the same nature. The simplest case is that already mentioned, where equivalent amounts of cathion and anion pass simultaneously into the un-electrified condition. This is, for instance, what takes place in the electrolysis of fused magnesium chloride; the anion here is chlorine, which passes from the ionic state (in which it is present in the fused salt) into that of ordinary unelectric chlorine, and it escapes as gas at the anode—the latter being usually made of carbon. An exactly similar process goes on with the magnesium ions at the cathode; these change into non-electric magnesium, *i.e.* into the ordinary metal, which separates there.

But the condition—that equal amounts of positive and negative electricity must be present at the same time in any given portion of the electrolyte—may be satisfied otherwise. If at the spot where, by the action of the current, a definite quantity of negative electricity must leave the liquid, an equal amount of positive electricity enters, the condition is likewise fulfilled, since it is a general law that any movement of positive electricity may be replaced by an equal and opposite movement of negative. The chemical reaction corresponding to this second case is, however, a totally different one, for now we have the anion remaining in the liquid and the equivalent amount of cathion leaving it. This result is attained when the electrodes are made of some substance which can pass into the ionic state when charged with positive electricity. Should, for example, the anode in the case mentioned above be made of iron, or any metal other than the noble ones, the chlorine will not leave the ionic condition, but on the contrary an equivalent amount of iron will pass into the state of the opposite ion.

The transformation of a metal into the corresponding cathion being termed oxidation and the opposite process reduction, the general statement may be put in these words—the anode effects an oxidising action and the cathode a reducing. This applies also for substances like chlorine and iodine which can pass from the non-electric state into that of *negative* ions; these two are oxidised at the anode and again reduced at the cathode.

Lastly, there is a third reaction possible at the electrodes. The necessary alteration in the amount of electricity may also be brought about by one ion passing into another which has a larger or smaller electric charge without its chemical composition undergoing any change. Such ions with different charges and correspondingly different valencies are found more especially among the metals; thus mercury and copper exist as mono- and divalent ions, tin di- and tetravalent, iron and chromium di- and trivalent, and thallium and gold mono- and trivalent. In accordance with this a ferrous salt changes at the anode into a ferric, and a mercuric salt at the cathode into a mercurous, if the conditions are such that the resulting ions of different valency can remain permanent.

Only a limited application of these various possibilities is made in electro-analysis, this being entirely confined for the present to the first-mentioned reaction, viz. the transformation of metallic ions into the non-electric state, or the separation of dissolved metallic elements in the form of a compact metal. The following conditions apply here :—

Every metal has a definite difference of potential with respect to the solution of any one of its salts, which, with given temperature, is alone dependent upon the concentration of the metallic ions in the solution. This difference may be either positive or negative, and the transformation of the metal into the ionic condition must accordingly follow either with the gain or the expenditure of energy. We have the first occurring in the case of the easily oxidisable metals, *i.e.* of the metals which pass readily into the ionic condition; in this group are comprised the metals at the so-called "positive" end of the electro-chemical series, from potassium to lead. On the other hand, the conversion of the metals succeeding lead—*i.e.* copper, mercury, silver, etc.—from the metallic to the ionic state requires the expenditure of energy, and the converse transition of ions into ordinary metal is accompanied by a gain of energy; these metals are thus spoken of as being easily reducible. If we were therefore to make a mixture of the electrolytes of all the metals and subject them to a gradually increasing electro-motive force, the metals would be separated in the order of the differences of potential existing between themselves and their electrolytes; the so-called noble metals would be the first to appear, and then—with a higher potential—copper, then lead, iron, tin, zinc, and so on.

If this mixture consisted merely of the metallic cathions and the requisite anions, the analysis might be carried down to potassium. But in an aqueous solution this point is never reached, because water contains the cathion hydrogen, and this is separated out at a definite potential. When the potential reaches the value requisite under the existing conditions for the separation of the hydrogen, the metallic ions which follow it in the series cannot become discharged, and all possibility of the electrolytic separation of these metals is at an end.

The position of hydrogen in the electro-chemical series is, however, by no means so definite as the positions of the solid metals, this being due to its gaseous state of aggregation. From this cause there arises the possibility of very exaggerated super-saturation phenomena, so that with suitable precautions hydrogen may be relegated to a position far behind that of the positive or zinc side of the series. As a matter of fact its place under normal conditions is near to that of lead, and in the electrolysis of the salts of all the more positive metals—*e.g.* of cadmium or zinc—we must obtain hydrogen from the aqueous solution of metallic salt instead of the metal itself. And this is really the case if the electrolysis is carried out with a very weak current, so that the resulting super-saturations have time to disappear. But if the strength of the current is increased, or to speak more correctly— the density of the current, *i.e.* the strength divided by the electrode surface, this reaction is checked, and we get for the most part an electrolysis of the metallic salt instead of an electrolysis of water. Further than zinc, however, it is scarcely possible to get in this way under normal conditions, although Bunsen has proved that even barium and other alkaline earth metals can be separated from warm concentrated solutions by using currents of excessive density. A mercury electrode is of especial value here, since the supersaturation of the hydrogen can attain a far higher value upon the smooth surface of the liquid metal than it can on ordinary solid electrodes.

As has been already mentioned, the potential at which the separation of any given metal takes place is dependent upon the concentration of the ions, and it must therefore be the further towards the zinc side, the less the concentration of the corresponding cathions. The difference is, however, not great for such concentrations as we have to deal with in analysis. The diminution of the ion concentration to the thousandth part of its original value (which forms the limit for most quantitative determinations) causes in the most extreme case—that of monovalent metals —a difference of potential of 0·17 volt, while in the case of a divalent metal the difference is only half as much. The differences existing between the various metals are for the most part much greater than this.

Quite other conditions prevail if the concentration of the metallic ion undergoes alteration through the latter passing into some complex compound, *i.e.* a compound in which it does not show the reactions of its ions. Although even in such a case we must assume that the solution contains a certain amount of the metal in the state of ordinary ions, this amount may be so extremely minute under certain circumstances as to be quite beyond the limits of analytical proof. The apparent electro-chemical position of the metal then undergoes marked displacement, which is invariably towards the zinc side—in other words, the metal behaves as if it were less "noble" than it really is. For example, gold is unaffected by the oxygen of the air, even when in contact with acids or other reagents, but it is attacked by a dilute solution of potassic cyanide when the air has access (there is no action in absence of air). The reason of this is that gold forms with cyanide of potassium a complex salt —potassium aurocyanide. In solutions of this salt the gold is present almost entirely as the group $Au(CN)_2$, and the concentration of the gold ions proper is so small that about the same difference of potential exists between the metal and the solution as between copper and hydrochloric acid (containing some copper); the oxygen of the air therefore acts upon the system so produced, as it does upon copper in hydrochloric acid, *i.e.* the gold is dissolved with absorption of oxygen.

The above exposition includes the points which are of most moment for the application of electrolysis to chemical analysis.. The great advantage of this procedure lies in the fact that, by the transformation of metallic ion into metal, the *mechanical separation* of the latter is effected without any further labour of filtration, etc. True, this advantage is only gained if the metal separates out in a compact mass, which is not the case under all circumstances; it is therefore essential for the practice of electro-analysis to learn the conditions under which the metal will assume the desired form. No general rule can be given for this as yet, so that we must be content for the present to find out empirically what the most favourable conditions are. Further, the separation takes place at a definite spot, *i.e.* at the cathode; the substance sought for can thus be compelled to betake itself from a dilute solution to a given point, whereby the handling of large quantities of liquid is avoided. Lastly, after the necessary arrangement of apparatus, etc., has once been made, the electrolytic process involves no labour or attention from beginning to end, and the accuracy of the analysis is thus far less dependent upon the skill of the operator than is the case in the ordinary mechanical methods.

The heavy metals constitute almost the only elements which have as yet been treated methodically in electro-analysis; the light metals take a position in the electro-chemical series which is too far apart from that of hydrogen to allow of their electrolytic separation from an aqueous solution. Most metals separate out at the cathode; it is, however, important to note—what Becquerel proved as long ago as 1830—that those metals which yield electrically conducting peroxides, more especially manganese and lead, can be separated very conveniently in this form at the anode.

The *quantitative separation* of the metals by electrolysis rests upon the differences of potential—just described—which are requisite for the separation in the metallic state. We may either prevent the separation of certain metals, which would come down under ordinary conditions, by converting them into complex compounds whose potential of precipitation lies very high, and this is the method which has been generally followed up to now. Or we can use from the beginning a *measured* electro-motive force, which is

H

higher than the potential necessary to precipitate the noblest among the metals present, but lower than the potential of decomposition of the succeeding metals. By taking advantage of the circumstance that the stability and therefore also the potential of decomposition of the complex salts formed by different metals under like conditions are often very different, we may frequently vary the conditions sufficiently to allow of our getting at those which are most favourable for the purpose.

CHAPTER V

THE QUANTITATIVE DETERMINATION OF SUBSTANCES

1. *General Considerations*

WITH the recognition of a substance—a point often involving the separation of its constituents from one another—the aim of qualitative analysis is achieved. If, however, we wish an answer to the further query,—how much of each different substance is present?—then a new problem awaits us, viz. the *measurement* of their quantities.

It does not follow that the measurement of a substance must always of necessity be preceded by its separation from everything else, any more than in the case of its recognition. It is a very common thing in quantitative analysis to have to estimate in a complex sample the amount of one constituent without paying any heed to the others present. We have thus to consider separately the methods which allow of quantitative estimation only after a previous separation, and those in which no such separation is required.

In order that a substance may be capable of convenient and accurate measurement, it must fulfil certain conditions. If, for instance, it has to be weighed, it must be unalterable in the air, non-deliquescent, and capable if possible of withstanding a red heat without undergoing chemical change. It is at once evident that only a comparatively small number of substances will satisfy those conditions. On this account the most extended use is made in analytical chemistry of the law of constant mass-proportions. A compound which is in itself unsuited for measurement is converted into some other possessing the desired properties, and its quantity is calculated from the measured quantity of the latter, in accordance with the law that the amount of the original substance stands in a constant ratio to that of the other into which it has been transformed. This ratio can be calculated from the law of combining weights, according to which all chemical combinations between any elements whatsoever take place in the ratio of definite relative numbers, those numbers being termed the atomic weights of the elements. The sum of the atomic weights of the compound is its combining weight, and the relation between the combining weights of two substances, one of which can be prepared from the other, is the ratio between the weight of the original amount of the first substance and the resulting amount of the second.

Given, therefore, a knowledge of the atomic weights of the elements and of the equation representing the chemical change in question, we can calculate the coefficient which reduces the amount of the original substance to that of the ultimate product, and *vice versâ*.

It is not necessary that any constituent of the original compound should be present in the one which is finally obtained. Thus the quantity of hydrochloric acid present in a solution may be estimated by allowing the latter to act upon an excess of carbonate of lime (marble), passing the carbon dioxide generated through a solution of baryta, and weighing the precipitated barium carbonate. In accordance with the equations

$$2HCl + CaCO_3 = CaCl_2 + H_2O + CO_2,$$
$$Ba(OH)_2 + CO_2 = BaCO_3 + H_2O,$$

we know that $1 BaCO_3$ is thus obtained from $2HCl$, and we can therefore calculate the required factor from the formula-weights $2HCl = 72.92$ and $BaCO_3 = 197.04$; $\frac{72.92}{197.04} = 0.3701$ is the coefficient by which the amount of barium carbonate obtained must be multiplied, in order to give us the original amount of hydrochloric acid.

And what holds good for weighing applies also to any other mode of estimation; hence we have a great variety of methods of measurement. We shall always assume, in what follows with respect to this, that use is made of the process of transformation if necessary. The chief point to be considered in the application of this is that any such transformation shall be easy and complete. When needful, the agreement between the theoretical factor and the empirical must be checked from the weighed quantity of the original substance; and any method which does not fulfil this condition —*i.e.* in which secondary reactions go on—is to be looked upon with suspicion and rejected, or at least only applied in default of a better one.

2. *Pure Substances*

When once substances have been separated from one another, the simplest and most reliable mode of determining their amounts is to weigh them. By means of the balance we determine directly the force with which the object weighed is attracted to the earth; and since we know that this force is proportional to the mass of the object, weighing becomes a determination of mass. The other properties which vary with the amount of substance, more especially the volume and the quantity of chemical energy, are on their part proportional to the mass, so that what we call the *quantity of a substance* is arrived at by weighing.

In place of weighing, we may measure other properties which are proportional to the mass. Prominent among these is the *volume*, whose measurement is often very much easier, and sometimes also more accurate than a weighing would be.

In the case of gases the measurement of the volume is, as a general rule, preferable to weighing, because the weight here constitutes but a small fraction of that of the containing vessel, and hence the influence of errors of weighing becomes very great. The effect of temperature and pressure on the volume of a gas is eliminated by always reducing to normal temperature and pressure (0°C. and 76 cm. mercury pressure), according to the formula

$$v_0 = \frac{pv}{76(1 + 0\cdot 00367 t)}.$$

The weight is got by multiplying the reduced volume v_0 by the weight of unit volume of the gas. In making rigorous calculations we have, however, to bear in mind that a column of mercury 76 cm. high is not a perfect definition of normal pressure, since the pressure thus determined depends further on the intensity of the force of gravity at any spot. In other words, the weight of the reduced unit volume of gas depends upon the geographical latitude and the height above sea-level. It would, therefore, be better to abandon the old definition of normal pressure, and to refer the latter to absolute units.

In determining the *quantity of a liquid* from its volume, it is only necessary to allow for the temperature, the compressibility of liquids and the alterations in atmospheric pressure being so small as to be negligible. For temperature there is no general law here; the expansion on warming must be determined separately for each individual substance. As a rule only the apparent expansion by heat is to be taken into account when measuring the volume of a liquid, *i.e.* the difference between the expansion of the liquid itself and that of the containing vessel. The volume is then reduced by the coefficient of apparent expansion to that for the temperature at which the density has been determined, or the densities at various temperatures may be estimated; the product of density and volume then gives the weight.

It may be said that a measurement of the volume of a solid is never applied in analysis for the purpose of arriving at its weight, since a solid cannot adapt itself to the filling up of an empty space as a liquid or a gas can do. The attempts which have been made from time to time to determine the amount of a precipitate without washing it, from the mean specific gravity of the precipitate plus liquid and the specific gravity of the latter alone, are based in principle on volume-determination. The inaccuracy of the results obtained has, however, prevented any practical application of this method; this arises from the density of a precipitate varying, and being influenced by adsorption in a way that cannot be controlled.

3. *Binary Mixtures*

Quantitative estimations in homogeneous binary mixtures (more especially in solution), the nature of both of whose constituents is known, are constantly carried out without a separation being necessary. To this end we only require to fix on any property which has a different value for the two constituents, and determine this in a sufficient number of mixtures of known composition to allow of its other values being interpolated; then from the observed value of this property in an unknown mixture, the composition of the latter can be arrived at.

The law, according to which the numerical values of the property in question depend upon the composition, need not for this purpose be given in strictly mathematical form. It is sufficient to collate the numbers empirically by a suitable interpolation-formula, or—more thoroughly and conveniently—by a curve, whose abscissæ indicate the quantity-ratios (*e.g.* in percentages of the total amount), and ordinates the numerical values of the specified property. In this way we get in general a curve, but in certain cases a straight line, connecting the ordinate-values of the pure substances together. The last case is an expression of the fact that in the process of mixing nothing has taken place to exercise any influence upon the individual properties of the two constituents; or, to put it differently, the property of the mixture is merely the sum of the properties of its constituents.

As already stated, such behaviour is rare in the case of mixtures of liquids, although universal with gases. The composition of a binary gaseous mixture can therefore be deduced from the measurement of any one definite property, for which the values in the pure constituents are known. Of all properties the specific gravity is the most serviceable from a practical point of view.

The determinations of the quantity-ratios from the measurement of a given property are the more accurate the more exact the measurement is in itself, and the greater the difference between the values appertaining to each of the two constituents. The most favourable case in this respect is the extreme limit, in which the value for the one constituent is zero, *i.e.* where only one of the constituents possesses the property in question. The measured property-value is then almost or altogether a measure of the relative amount of the substance possessing that property. Among such properties, which may be termed *special*, in contradistinction to the other *general* properties, may be instanced the rotation of the plane of polarisation of light, the colour, and the electric conductivity, etc.; all of these lend themselves in an especial manner to quantitative determinations, and are largely applied for this purpose.

As already mentioned, the general properties, which possess a finite value for all substances, do not allow of such an accurate quantitative determination under otherwise similar conditions, because in them the measure of the amount is approximately or exactly proportional only to the *difference* between the values for the mixture and that for the pure constituent. Notwithstanding this these properties are very largely made use of, on account of the ease and accuracy with which they can be measured.

Of the general properties the most prominent is the specific gravity, which can be measured with extreme accuracy by the specific gravity bottle, and very quickly and easily by the aræometer. Since the case of the property being purely additive (cf. p. 104) hardly ever occurs here (aqueous solutions in particular show large deviations), a series of measurements has to be carried through for each individual substance beforehand, so as to allow of the requisite interpolations being made. Temperature has a very appreciable influence on specific gravity, and must consequently be always allowed for; it is therefore necessary to work at a perfectly definite temperature.

Among other general properties the refraction-coefficient may be mentioned, since it can be as widely utilised as specific gravity itself. Its measurement is, however, less easy to carry out as a rule, or else less exact. Other auxiliary aids are boiling point or vapour pressure, melting point, expansion coefficient, internal friction, etc.

4. *Indirect Analysis*

In addition to the physical procedure for determining the quantities of substances in binary mixtures, a procedure depending upon the measurement of properties, there is the chemical one by which, after the weight of the mixture has been noted, the latter is transformed either into another mixture or into some homogeneous substance. From the change in weight thus brought about, the composition of the mixture may be deduced in a way similar to that already explained, the relation between the change in weight and the composition being a simple linear one, since weight is a purely additive property.

Suppose, for example, that we have a mixture of the chlorides of potassium and sodium, we can arrive at its composition by transforming the chlorides into sulphates. We calculate from the atomic weights that 1 grm. of sodium chloride will yield 1·2147 grm. of sulphate, while a grm. of potassium chloride will only yield 1·1683 grm. of sulphate; a mixture of the two salts must therefore give a value lying between those two figures. If this value is 1·2015, then $\frac{1\cdot2147 - 1\cdot2015}{1\cdot2147 - 1\cdot1683} = 0\cdot285$ grm. will represent the proportion of chloride of potassium which is present.

Numerous schemes of indirect analysis can be drawn up on the same principle. But the process, although convenient, has the drawback of multiplying the error of experiment in a greater or less degree. In the above example, which is an unfavourable one, the whole difference in the final weight must be something less than 0·0464 grm. for every gramme of the original mixture taken, consequently any error of weighing becomes multiplied 22 times in the result. When, therefore, we have occasion to make use of indirect analysis, we must choose our process so that there shall be the largest possible weight-difference between the transformation products of the two separate constituents.

In place of direct weighing, the quantities of substances may be determined by physical or chemical methods; but the principle of the procedure remains the same.

5. *Tertiary Mixtures, etc.*

Quantitative estimations in complex mixtures may be carried out without any preliminary separation, if the substance to be estimated possesses some *special property*, from whose measurement its quantity can be deduced. But before applying any such procedure an investigation must first be made to see that the relation between the value of the specified property and the amount of the substance is not affected in any way by the presence of other bodies. It would, for instance, be utterly fallacious to try to arrive at the percentage of sodium chloride present in an aqueous-alcoholic solution from the electric conductivity of the latter, if in our calculation we referred the conductivity found to a table constructed from results given by purely aqueous solutions. For although alcohol is a non-conductor itself, it exerts an influence on the conductivity of solutions to which it is added, so that the relation between this property and the amount of substance present becomes quite altered.

It is only when no such influence is exercised by any of the other substances present that this procedure can be followed with advantage. For although it may be possible to determine the influence of the foreign body and to tabulate it, the use of any such table (apart from the additional work required to construct it) necessitates a knowledge of the quantity of the foreign substance present, and therefore very often a special analysis for this purpose.

Cases in which the procedure is practically applicable are consequently not very common, and, since we almost never find the special property in question to be *absolutely* independent of outside substances, the methods are not very exact. The determination of cane sugar from the optical rotatory power of its solution and the various colorimetric analyses may be taken as examples. Serious errors have been repeatedly brought to light in the last-mentioned of those cases, which have arisen from taking for granted without sufficient proof that the foreign bodies present exercised no influence, when as a matter of fact their influence was a material one.

6. *Methods of Titration*

Chemical methods for the estimation of one constituent in a complex mixture are far more reliable and also far more numerous than physical ones. They are based upon the principle of subjecting the substance in question to some chemical reaction by the addition of a suitable reagent, whereby either the complete conversion of the original substance or the slightest excess of added reagent can be at once recognised by some striking sign. It is as a rule much easier here to judge and also to check whether, through any other possible chemical reactions, the condition has been infringed—that no other substance present shall exert an influence on the determination; hence such methods are capable of unusually wide application. The quantitative estimation depends in this case upon the measurement of the quantity of reagent which must be added, in order to completely transform the substance under examination. This quantity is most conveniently arrived at by noting the volume of the reagent-solution used, the strength being also of course known. This is not, however, an essential characteristic of the method, for, when exceptionally accurate determinations are required, it is not unusual to replace the measurement of the volume of the reagent by weighing, which, in contrast to the volumetric measurement, is unaffected by variations of temperature.

Methods of titration may be divided into two groups, viz. those in which the disappearance of the original substance furnishes the end-phenomenon, and those in which the excess of reagent performs this duty. The methods of the second group are in their turn capable of further division, seeing that in some cases an excess of the reagent is at once directly apparent, while in others this excess is only rendered visible through the aid of an added medium—the *indicator*.

As an example of the first group we may take iodometric analysis. By the reaction

$$I_2 + 2Na_2S_2O_3 = 2NaI + Na_2S_4O_6$$

the dark-coloured free iodine is transformed into sodium iodide, or, more correctly, into colourless iodine ions. One has therefore to go on adding thiosulphate solution of known strength until the yellowish-brown colour of the free iodine has vanished. The end-point of this conversion is, however, far easier to see if, towards the close of the titration, any iodine still present is converted into blue iodide of starch, by the addition of a few drops of clear starch solution.

The estimation of iron by permanganate of potash is a good example of the first section of group 2. So long as any ferrous salt remains in the solution, the added permanganate is at once changed into colourless compounds (manganous and potassium salts). But, when all the ferrous salt gets used up and this conversion consequently ceases, the pink permanganate colour remains permanent, thus showing the end-point of the analysis.

This method is obviously only applicable when the reagent possesses some striking characteristic (in the above case the pink colour), which disappears through the reaction. Should this be wanting, an indicator must be used.

The typical example of the indicator-method is the process of acidimetry or alkalimetry, by means of which the amount of base or acid—or, to be more precise, the amount of hydroxyl or acid-hydrogen—is estimated in a solution. Since these substances give no indication of their presence by any direct sign, a dye like litmus is added, whose colour depends upon whether the solution contains an excess of acid or of alkali. Thus, litmus gives with alkali a blue salt, which is decomposed by the slightest excess of acid with the liberation of the free litmus acid, a compound of reddish colour.

In place of a change in colour, the production or disappearance of a precipitate or any other striking phenomenon may be made use of. If the indicator cannot be actually employed *in* the solution which is being examined, some drops of it are spotted over a suitable surface (a white porcelain plate in colour reactions) and, after each successive addition of reagent, a minute quantity of the solution is brought into contact with one of these drops, until the reaction, which is characteristic of an excess of the reagent, sets in.

Since the time of Friedrich Mohr, the strength of volumetric reagents has been so regulated that an equivalent of the reagent in grammes is contained in a litre of the liquid, or in some multiple of a litre. The number of cubic centimetres of reagent used thus gives the amount of the substance under examination corresponding to the reaction-formula in milligramme-equivalents or some sub-multiple of this, and so the labour of subsequent calculation is reduced to a minimum. In cases, however, where a very large number of analyses of the same kind have to be carried out, more especially in technical working, the titrating solution is so made that 1 c.cm. of it corresponds to 1, 10, or 100 m.grms., or any other round figure of the substance to be determined.

PART II
APPLICATIONS

INTRODUCTION

IN order to give a living interest to the laws and principles which have been detailed in the first part of this book, and to illustrate the mode in which they are applied, I shall discuss in the second part the analytical properties of a number of different substances separately. My object in doing this will not be to attempt to teach analytical chemistry to beginners, for I am well aware that an appreciable time must elapse before the new views enunciated here can find general acceptance, and can exercise an influence upon the instruction of those who are commencing the study of the science. Students who read this book will thus do so not so much with the object of actually learning analytical chemistry from it, as of pondering over the scientific principles which underlie what they have already been taught by actual practice, so as to be able to apply this knowledge with greater freedom and certainty. It would of course be superfluous to attempt the treatment of every known substance, but it will be my endeavour to bring together here examples of all the cases which are typical and characteristic.

The division of the subject will be according to the usual analytical groups. I should like here to lay stress upon the treatment of the subject with respect to the ionic state, which the element sought for may assume, this point being essentially new, and, in my opinion, capable of direct application in teaching. If we adhere constantly to the point of view that analytical reactions are with very few exceptions *reactions of ions*,[1] then a review of the facts of analytical chemistry becomes at once infinitely simpler, and appeals in its practical bearing even to those who look upon the theory of electrolytic dissociation as a suspicious and blameworthy innovation.

[1] *Zeitschrift für physikalische Chemie*, 3, 596 (1889).

CHAPTER VI

THE HYDROGEN AND HYDROXYL IONS

1. *Acids and Bases*

COMPOUNDS whose aqueous solutions contain the hydrogen ion are termed acids, and those which contain the hydroxyl ion bases. They are recognised qualitatively by the colour reactions which they bring about in certain dyes, those reactions being utilised as indicators in the quantitative estimation of the hydrogen ion or the acids, and in that of the hydroxyl ion or the bases.

(If a dye is to be of any use as an indicator, it must be either acid or basic in its nature, and must have a different colour when non-dissociated from what it has in the ionic condition.) Further, it must not be a strong acid or base, because it would then break up into its ions while in the free state, and would thus show no change of colour on neutralisation. For, when a strong acid is neutralised, only its free hydrogen ions form water with the hydroxyl of the base, the anion remaining unaltered. A weak acid on the other hand exists for the most part as undissociated molecule in solution and not as ion, the ions being produced only after neutralisation, *i.e.* after the conversion of the acid into neutral salt, seeing that the neutral salts of even weak acids undergo very complete ionisation.

2. *The Theory of Indicators*

For the rest, the properties of an indicator depend mainly upon the extent to which it is dissociable. If it is a very weak acid (and precisely analogous considerations hold good for basic indicators), then acids of moderate or even small ionisation, if present in the slightest excess, will give up their hydrogen to it, thus producing the colour-phenomenon which accompanies the change from the ionic state to that of the non-ionised molecule. Such indicators will thus be sensitive, and capable of being used for the measurement of tolerably weak acids like acetic. They can only be employed with strong bases, however, as with weak ones they form salts incompletely, the latter being decomposed by hydrolysis with the water present (cf. p. 66); with weak bases, therefore, they give the colour-phenomenon due to the formation of ions very imperfectly, and the change is not a sharp one.

Phenol-phthaleïn is a good example of a very weakly acid indicator, which is colourless in the molecular state, but intensely red when dissociated into ions. The solution which is coloured red by alkali contains the salt of phenol-phthaleïn in the ionic state, and becomes colourless after neutralisation and addition of the faintest excess of acid, from the formation of the colourless non-ionised molecule. Ammonia is, however, too weak a base in itself to yield a normal salt with phenol-phthaleïn in very dilute solution; in other words, to give the free ions of the salt. In fact a distinct excess of ammonia is required to overcome the hydrolytic action of the water, hence the change of colour in presence of salts of ammonium is anything but sharp, and only apparent when a considerable excess of base is present. For *acidimetry*, especially for the titration of weak acids, when one can choose any base for neutralisation that one likes, and therefore fixes on a strong one (baryta water is best), phenol-phthaleïn is an admirable indicator; but it is unsuited to alkalimetry, since its use must be restricted to the very strong bases.

Of the other well-known indicators methyl-orange stands in the opposite category. It is an acid of medium strength, its ions being yellow while the non-dissociated molecule is red. The pure aqueous solution of the acid ionises of itself in a marked degree and therefore shows a mixed colour; but the addition of a trace of a stronger acid lessens this ionisation, in virtue of the mass-action of the hydrogen ions (p. 64), and then the colour of the undecomposed molecule predominates.

When, however, methyl-orange is added to a basic liquid, the salt is formed and we get the yellow colour of the ions. If we now neutralise with a strong acid, the reaction mentioned in the preceding paragraph as due to the presence of excess of hydrogen ions takes place, and the colour is reversed. But should the acid be weak, *i.e.* only slightly ionisable (the ionisation being still further reduced by that of the neutral salt formed in the liquid), the quantity of hydrogen ions on passing the point of neutralisation is too small to allow of the formation of a visible amount of non-ionised molecules of methyl-orange, and the red colour only appears after a very considerable excess has been added, and then only by degrees; in other words, the reaction is not a sharp one. Methyl-orange is therefore not suited to the titration of every acid.

But when it is a question of the titration of bases, even of weak bases, methyl-orange is the proper indicator to use, for its strongly acid nature permits of its forming salts with very weak bases, and those salts are not hydrolysed to any extent by water; it thus gives sharp end-reactions in cases where indicators of slightly acid nature are useless.

The other acid indicators lie between these two extremes, and the conditions under which they are applicable can therefore be judged of from what has just been said.

Considerations of precisely the same kind apply to basic indicators. For the titration of weak acids a strongly ionised indicator is alone of use, while weak bases require as weakly basic indicators as possible.

But neither here nor in the case of acid indicators must we go to the extreme of ionisation. For an indicator which is as much ionised as the strongest acids (hydrochloric, nitric, etc.) *will show no alterations of colour at all* in acid and alkaline solutions. In an acid solution it is already ionised to its full extent, practically speaking; its anions are therefore present in the free state, and do not first become free when the salt is formed. Further, since the anions undergo no change through the neutralisation of the liquid, there can be no alteration in colour. Examples in point are to be found in (the strong) picric and permanganic acids, which show the same colour in acid as in alkaline solution.

3. *The Presence of Carbonic Acid*

Certain difficulties arise in acidimetry from the circumstance that basic solutions absorb carbon dioxide from the air, whereby their titre becomes altered. This source of error must be rigorously excluded in the determination of weak acids. In such cases, therefore, the alkaline solution must be carefully guarded from atmospheric carbonic acid (*e.g.* by interposing tubes filled with soda-lime), and baryta water is the best alkaline liquid to use, seeing that it cannot retain carbon dioxide in solution, besides acting much less on glass than solutions of potash and soda do.

On the other hand, when strong acids have to be estimated, the action of carbonic—itself a very weak acid—can be set aside by using for indicator an acid of medium strength. Methyl-orange is the best here, acting satisfactorily, not merely in the titration of alkali which contains carbonate, but in that of a carbonate itself; and, further, the end reaction is sharper, the more concentrated is the solution. The correctness of this statement can be verified by a careful study of the conditions of equilibrium in point, but it will be sufficient merely to refer now to the fact that, with increasing dilution, all acids approximate to one another in their degree of ionisation; hence the differences in ionisation, upon which the process is based, become gradually eliminated as the solutions become more dilute.

Hydrosulphuric acid acts similarly to carbonic, only its ionisation-constant is somewhat greater and its acid properties are therefore rather more marked.

4. *Polybasic Acids*

While monobasic acids allow of a sharp titration, even when they are relatively weak, some polybasic acids of markedly acid character show the change of colour very gradually,—an indication of the hydrolysis of their neutral salts.

The cause of this striking phenomenon, for which sulphurous and ortho-phosphoric acids may be taken as examples, is the *gradual ionisation* already referred to on p. 61, according to which the different hydrogen atoms of polybasic acids pass at very different rates into the ionised condition—the second more slowly than the first, and the third more slowly than the second, and so on. So far, however, as the change in colour of the indicator is concerned, it is only the last—*i.e.* the weakest—hydrogen atom that has to be taken into account, since the first (and second, etc.) have been already accounted for by the first portions of added base. If the ionisation-constant for this hydrogen atom is very small, hydrolysis of the corresponding salt in the aqueous solution ensues (cf. p. 66), and the result is an indistinct colour-reaction as explained on p. 118.

The different behaviour of polybasic acids with different indicators depends upon the same thing. Phosphoric acid comports itself with methyl-orange as a monobasic acid, *i.e.* only its first hydrogen atom is sufficiently ionised to give up to the yellow ions the hydrogen necessary for the formation of the red non-ionised molecule. With phenol-phthaleïn, which is a very much weaker acid, phosphoric acid behaves on the other hand, as if it were dibasic, since this indicator requires a much smaller concentration of the hydrogen ions for its conversion into the colourless non-ionised compound. The third hydrogen atom of phosphoric acid, lastly, is one of so weak an acid that the corresponding alkaline salt becomes hydrolysed to a very considerable extent in aqueous solution, so that a titration is impossible.

The fact that carbonic acid can be titrated as a monobasic acid with phenol-phthaleïn for indicator, is explicable in the same way.

Since we are dependent upon comparatively small differences of ionisation in all these processes, the colour changes are less sharp than in the case of strong monobasic acids, and the conditions of the reaction are somewhat disarranged by dilution. If estimations of the kind must be made (and this is not to be recommended as a rule), the solution to be titrated should be kept as concentrated as possible.

Attempts have been made to refer the phenomena which have just been described to the unsymmetrical constitution of the acids in question. But they occur with many acids about whose symmetrical constitution there is no doubt, and at the same time are wanting in the case of others which are as certainly unsymmetric. The causes which determine a greater or lesser difference in the consecutive coefficients are partly known, but they cannot be entered into here.

CHAPTER VII

THE METALS OF THE ALKALIES

1. *General Properties*

THE metals potassium, rubidium, cæsium, sodium and lithium, are invariably present in solutions as monovalent positive ions, and never as complex ones. They therefore always show the reactions of those ions, and never any so-called anomalous reactions. The hydroxides are readily soluble in water and are very completely ionised, so that they constitute the strongest bases known. With the ordinary precipitating reagents they yield only soluble salts, remaining in solution after any other metals present have been thrown down; hence advantage is taken of this to separate them from the latter.

2. *Potassium, Rubidium and Cæsium*

The alkali metals form sparingly soluble salts with hydro-silicofluoric and hydro-platinichloric acids. The first of these precipitates potassium, rubidium and cæsium ions as well as those of sodium, and cannot therefore be used for separations. Hydro-platinichloric acid gives difficultly soluble salts of the formula M_2PtCl_6 with potassium, rubidium and cæsium, and easily soluble ones with sodium and lithium. When potassium and sodium have to be separated from one another, the chlorides are evaporated with excess of hydro-platinichloric acid to a syrup, and the residue allowed to stand covered with excess of alcohol, which dissolves the sodium platinichloride readily. As both sodium chloride and anhydrous sodium platinichloride are but slightly soluble in alcohol, we must always have an excess of hydro-platinichloric acid present, and must also take care that the residue from evaporation on the water-bath is never made completely dry. The platinichloride of potassium, dried at 110°, still contains traces of water, so that its apparent weight is a little greater than the reality.

No analytical method is known for the separation of potassium, rubidium and cæsium; use is made of the different solubilities of their platinichlorides or acid tartrates to effect a "fractionation" or approximate separation. For an actual determination only an indirect process can be followed, e.g. the metals are first weighed as chlorides and then as platinichlorides; it is of course an essential condition here that only two of the elements shall be present.

Potassium may also be precipitated as acid tartrate by the addition of tartaric acid. Since the free acid is thereby produced from the potassium salt, and exerts—if it be concentrated—a solvent action upon the precipitate (cf. p. 81), i.e. hinders the formation of the latter, it must either be rendered innocuous by the addition of sodium acetate (p. 65), or a solution of sodium-hydrogen tartrate must be used for precipitating instead of tartaric acid itself; by this means the production of free acid is avoided. The second of these two methods is to be preferred, inasmuch as it permits of more tartaric acid ions being brought into solution than is possible when the (slightly ionised) free tartaric acid is employed, and thus diminishes more effectively the solubility of the resulting hydrogen-potassium tartrate.

This salt, potassium-hydrogen tartrate, shows the phenomena of super-saturation in a very marked degree; hence the reaction just detailed must be carried out in as concentrated a solution as possible, and the precipitate must be allowed to remain in the liquid for a long time, with frequent shaking.

The flame-reaction serves as a qualitative test for potassium. The potassium flame contains violet and red rays, and appears reddish when looked at through cobalt-blue glass, since it is the red rays that are chiefly transmitted. The yellow sodium light, which hides that of potassium from the naked eye even when but minute quantities of sodium are present, is completely stopped by cobalt glass, so that the yellow flame proceeding from a mixture of potassium and sodium compounds appears red when viewed through the latter, while a pure sodium flame is invisible.

With such a mixed flame the spectroscope shows the red lines of potassium, and also the violet lines faintly, together with the double yellow sodium line.

3. *Sodium*

Sodium is determined quantitatively in a mixture by first weighing the chlorides of sodium and potassium together, then estimating the potassium as platinichloride, and deducting the chloride of potassium thus formed from the original total amount; the difference gives the chloride of sodium.

The qualitative test for sodium is the yellow flame coloration. Since, however, compounds of sodium are distributed everywhere in nature, and this flame reaction is excessively delicate, one has to note here the length of time that the yellow shows. Traces of sodium, arising, *e.g.*, from the presence of dust or from the fingers touching the platinum wire, give but a transitory coloration, while measurable amounts of sodium compounds show the phenomenon for several minutes.

4. *Lithium*

Lithium gives an intensely red flame, which shows a red and an orange yellow line when analysed by the spectroscope. The lithium light is of smaller wavelength than that of potassium, and is absorbed by cobalt glass.

The reactions of lithium are more like those of the alkaline earth metals than of the metals of the alkalies. It forms a sparingly soluble carbonate and phosphate, its chloride—unlike those of the alkalies—is soluble in an anhydrous mixture of alcohol and ether, and it becomes alkaline when heated strongly in moist air, like the chlorides of calcium and magnesium. It does not give a precipitate either with hydro-platinichloric or with tartaric acid.

Lithium is estimated quantitatively as phosphate, Li_3PO_4, by precipitating the solution with tri-sodium phosphate (*i.e.* with a mixture of common sodium phosphate, Na_2HPO_4, and caustic soda).

5. *Ammonia*

The salts formed by the combination of ammonia with the acids contain the ion $NH_4^{\bullet}$, which behaves in many respects like the potassium ion. Like the latter it forms a sparingly soluble platinichloride and acid tartrate, and many of its salts are isomorphous with the corresponding salts of potassium.

Ammonia dissolves in water to ammonium hydroxide, NH_4OH, which becomes ionised to some extent. Its ionisation-constant in the formula $\frac{a^2}{(1-a)v} = K$, when v is expressed in litres is :— $K = 0.000023$; in its decinormal solution the ionisation amounts to 1·5 per cent. Ammonia therefore belongs to the weaker bases.

When its aqueous solution is warmed, the ammonium hydroxide changes into its anhydride ammonia, which partly escapes. All the ammonia can be driven out of a water solution by boiling; each escaping bubble of vapour creates a vacuum for the ammonia, in which the partial pressure of the latter is at first zero, so that the gas quickly diffuses out of the liquid and becomes expelled. Upon this the estimation of ammonia in its salts depends. These are distilled with a stronger base, and the escaping ammonia is condensed in some acid. For a quantitative determination a measured volume of acid of known strength is used, the excess of acid remaining after the operation being titrated with baryta water, with methyl-orange as indicator.

Ammonia possesses in a remarkable degree the power of forming complex ions of the general formula

$M + nNH_3$ with the elements, more especially with metals, these ions having often the same valency as the metallic ions themselves. The most stable compounds of this type do not show either the reactions of the metal or those of ammonia. The complexes of this kind are, however, of every degree of stability. The more readily decomposable among them are the most stable when in the form of salts; the free bases break up more easily into metallic hydroxide and ammonia. Most of them give off ammonia more or less quickly when heated alone, while they are all completely decomposed upon heating with caustic alkali or soda-lime, the nitrogen present escaping as ammonia.

A characteristic compound, dimercur-ammonium iodide, $NHg_2I + H_2O$, is formed as a yellowish-brown precipitate when even excessively attenuated solutions of ammonium compounds are added to an alkaline solution of potassium mercuriodide, K_2HgI_4 (Nessler's reagent). The mercury salt must be present in excess, or else more soluble mercur-ammonium compounds (poorer in mercury) are produced, and the liquid must also be rather strongly alkaline. Since we have to do here with the formation of a compound substance and not with a simple ion-reaction, the mixed liquids require to stand for some time before the process completes itself.

CHAPTER VIII

THE METALS OF THE ALKALINE EARTHS

1. *General Properties*

THE five metals calcium, strontium, barium, magnesium and beryllium form divalent positive ions; the three first can only exist in this state (in solution), while magnesium and beryllium are capable of uniting with ammonia to complex ions, though these are not by any means stable. Calcium, strontium and barium also yield more or less soluble strong bases with water, which are ionised almost as much as the alkaline hydroxides. Little, however, can be said with respect to the ionisation of the other two hydroxides, since they are so slightly soluble in water; but the behaviour of the salts enables us to conclude that magnesia is a moderately strong base and beryllia a weak one, the salts of the latter showing an acid reaction, and therefore undergoing hydrolytic decomposition by the solvent water. Speaking generally, beryllium—the metal of smallest atomic weight—acts as a connecting link with the trivalent metals of the next group, just as lithium — the alkali metal of smallest atomic weight—shows certain points of resemblance to the above divalent metals.

All the metals of this group give sparingly soluble carbonates and phosphates; and the three first yield sulphates whose insolubility increases in the order given. Of the sulphides of calcium, strontium and barium, that of barium is the most soluble and that of calcium the least, but they are all decomposed hydrolytically, since the monovalent ions SH' and OH', together with the metallic cathion, are present in the solution, instead of the divalent sulphur ion, S''; if the amount of water in the solution is carefully regulated, the hydroxides can be made to crystallise out. The sulphides of magnesium and beryllium undergo this hydrolysis to such an extent that sulphuretted hydrogen escapes, and the sparingly soluble hydroxides are precipitated.

2. *Calcium.*

Calcium salts are precipitated by carbonates, phosphates and oxalates. Carbonate of calcium comes down amorphous at first, and in that state is perceptibly soluble in water; but on standing, and more quickly upon warming, the precipitate becomes crystalline (assuming the rhombohedral forms of calc spar), and at the same time much more insoluble. This crystalline precipitate dissolves readily even in weak acids, and also when boiled with solutions of ammonium salts—*e.g.* the chloride—carbonate of ammonium escaping. Amorphous carbonate of calcium being more soluble, it is taken up even by a cold solution of sal ammoniac; hence it follows that calcium salts are not precipitated by carbonates when a sufficiency of ammonium salt is present.

Calcium oxalate is a very sparingly soluble compound. Free oxalic acid only precipitates the calcium salts of strong acids imperfectly, as the oxalate is soluble in free hydrochloric or nitric acid. Oxalate of ammonium, on the other hand, effects a complete precipitation, even in the presence of free acetic acid; the latter indeed has a certain solvent action upon pure calcium oxalate, but in presence of acetate and excess of oxalate this solubility diminishes almost to the vanishing point, the former lessening the solvent effect of the acid, and the latter the solubility of the calcium oxalate.

In a quantitative estimation the precipitated oxalate of calcium is either gently ignited and weighed as carbonate, or strongly ignited and weighed as oxide, the second procedure being the better. Precipitation as oxalate also serves as the qualitative test for calcium salts, after those of barium and strontium have been already removed.

Ammonia does not give a precipitate with the salts of calcium, being too weak a base; but potash and soda, especially if somewhat concentrated, throw down the sparingly soluble calcium hydroxide. Pure water dissolves this hydroxide to the extent of about 1 part in 500, but if an alkali is present the solubility rapidly diminishes, on account of the increase of the one ion hydroxyl, so that in a 10 per cent alkali solution lime is practically insoluble. This circumstance is important for the manufacture of caustic alkali by boiling alkaline carbonate with lime.

To the *Bunsen* flame calcium salts give a brick-red coloration, especially after moistening with hydrochloric acid; the spectrum is rather a complex one.

3. *Strontium.*

Strontium is precipitated as sulphate; it is necessary to add alcohol here when a more perfect precipitation is wanted, and an excess of the soluble sulphate employed is also advantageous. Sulphuric acid being less dissociated than hydrochloric or nitric, the two latter exert a distinct solvent action upon the sparingly soluble sulphates, since they induce the formation of non-ionised sulphuric acid (cf. p. 81). This phenomenon is, of course, most apparent in the case of calcium sulphate, which is more soluble than either of the other two alkaline earth sulphates,— gypsum in fact dissolves readily enough in hydrochloric acid. But the same thing applies in a lesser degree to strontium sulphate also, and it is therefore wise to avoid using any excess of strong acid in precipitating this salt, *i.e.* to keep the solution either neutral or acidified only by acetic acid.

Strontium sulphate can be readily and completely converted into carbonate by digestion with soluble carbonates. The laws which regulate such transformations can be easily deduced from the general law of equilibrium. Such conversions are always reciprocal, for, just as the above sulphate is changed into carbonate by a soluble carbonate, so can the carbonate be transformed into a sulphate by a soluble sulphate. There must, therefore, be some definite ratio between the ions SO_4 and CO_3 when neither of the two transformations can take place; and this ratio is necessarily that at which the two sparingly soluble salts both dissolve at one and the same time in water. For it is obvious that under such circumstances no mutual interaction can set in, and the concentrations of the ions SO_4 and CO_3 must stand in the ratio of the solubility-products, seeing that the amount of the Sr-ions, which is a factor in both products, is the same for each. The latter also applies for the case of soluble carbonates and sulphates being present together, consequently the ions SO_4 and CO_3 must stand in the same ratio here also.

It follows from what has just been said, that a solution containing excess of carbonate can have no action upon the solid carbonate; neither can one containing an excess of sulphate act upon a solid sulphate. If in this latter case solid carbonate should be present at the same time, it will continue to undergo transformation until the critical ratio of the two ions has established itself in the solution. Neither the absolute quantities of the solids nor the proportion existing between them has anything to do with the matter.

In the case of strontium the solubilities which we have to take into account are very different, the sulphate being considerably more soluble than the carbonate; the transformation of the former thus goes on much more quickly than that of the latter, and the sulphate must preponderate distinctly in the solution if there is to be equilibrium. With barium the solubilities are about the same, and therefore also the ratio of the two soluble salts in the state of equilibrium.

From dilute solutions strontium sulphate is not thrown down at the moment, a point of which use can be made to distinguish strontium from barium, a very dilute solution of a sulphate being of course employed as a precipitant; a saturated solution of gypsum is a convenient strength. We should be wrong in assuming, however, that it is the actual formation of sulphate which goes on so comparatively slowly; on the contrary, the latter is formed at once, as is shown by the measurement of the electric conductivity on mixing dilute solutions of strontium hydroxide and sulphuric acid together. The cause of the tardy precipitation is an ordinary super-saturation phenomenon.

In the *Bunsen* flame strontium gives a crimson coloration, the spectrum being like that of calcium somewhat complex; a blue line in the latter is specially characteristic. The chloride is the best strontium compound to use for the purpose, the spectrum indeed only showing in some cases after the test-substance has been moistened with hydrochloric acid.

4. *Barium*

Of all the sulphates of the alkaline earth metals, that of barium is the most insoluble. It therefore serves generally for the recognition and separation of the ion SO_4, in whose presence barium salts give rise to a fine powdery white precipitate. This precipitate is scarcely more soluble in dilute solutions of acids—even of strong ones—than in pure water alone. Since barium forms no complex ion, there is no aqueous solvent for the sulphate, which may therefore be considered the most insoluble of all precipitates.

In order to separate barium from strontium (whose recognition it hinders), it is either precipitated by hydro-silicofluoric acid, which does not throw down strontium, or by a neutral soluble chromate. Hydro-silicofluoric being a tolerably strong acid, its barium salt is but little more soluble in dilute acids than in water; for the opposite reason barium chromate dissolves in strong acids, hence its precipitation must take place either in a neutral or an acetic acid solution.

The two elements may also be separated by applying the laws for the equilibrium of soluble and insoluble salts which have just been explained. A solution which contains approximately equal equivalents of soluble carbonate and sulphate has no effect upon sulphate of barium, while it readily transforms sulphate of strontium into the carbonate. If then both metals have been precipitated as sulphate, the precipitate can be converted into a mixture of barium sulphate and strontium carbonate by digestion with the solution named, and then the strontium carbonate dissolved out by hydrochloric acid.

Barium salts impart a greenish colour to the *Bunsen* flame, and the spectrum is found to consist of a large number of bands (not lines).

5. *Magnesium*

Besides the normal divalent ions, magnesium can form complex ions with ammonia, and it therefore shows reactions in alkaline solution, in the presence of ammoniacal salts, which differ from those of the other alkaline earth metals. In acid solution it resembles the others for the most part.

Magnesium hydroxide is a distinctly weaker base than the other hydroxides of this group; it is incapable of forming the normal carbonate with carbon dioxide in presence of water, the hydrolytic action of the latter giving rise to a mixture of carbonate and hydroxide. On precipitating in the cold, soluble bicarbonate remains in solution in large quantity, and only separates out upon warming.

The hydroxide of magnesium is indeed soluble enough to turn red litmus paper blue, but becomes so sparingly soluble in the presence of excess of alkali that this can be made use of for its quantitative separation.

If to a solution of magnesium salt an excess of ammonia is added, the hydroxide only separates partially, owing to the formation of the complex magnes-ammonium ion, $Mg(NH_3)_2$. The hydroxide of this latter is unstable, and even the neutral salts are only permanent in presence of excess of ammonia. Hence ammonia yields no precipitate if an excess of ammoniacal salt has been previously added. Even caustic alkalies fail to produce a precipitate under those conditions, *i.e.* so long as sufficient ammonium salt is present; an excess of alkali, however, drives out the ammonia, especially on boiling, and then the precipitation is complete.

Magnesium is best estimated by precipitating the ammoniacal solution with the phosphate of an alkali metal, when ammonium-magnesium phosphate is thrown down. This precipitate being decidedly soluble in water, dilute ammonia has to be used in washing it; for, since ammonia is a dissociation-product of the precipitate, the latter is less soluble in its presence than in pure water alone.

Salts of magnesium give no coloration to the *Bunsen* flame.

6. *Appendix*

Aluminium.—The trivalent ion of aluminium has only a weakly basic character. All its salts show an acid reaction, and those with weak acids break up—when their solutions are boiled—into basic salts which separate out, and free acid which remains in solution. The hydroxide has no action upon litmus paper.

Aluminium hydroxide behaves with respect to the soluble bases in exactly the opposite way that hydroxide of magnesium does, for it is insoluble in ammonia but soluble in potash and soda. The solubility in these arises from its being able to act as an acid, the ions of which are $3H^{\bullet}$ and AlO_3'''; in the formation of the latter the Al-ions are used up, and consequently the hydroxide must pass into solution.

There is a point in which aluminium differs from the metals which have been already considered, but which we shall find repeated in most of those that have still to be spoken of, viz. the precipitation of the hydroxide is prevented by the presence of non-volatile organic acids. The cause of this in every one of these cases is the formation of complex compounds through the entrance of the metal into the hydroxyl of the acid. For, the non-volatile organic acids which show this action, all contain hydroxyl; and that the hydroxyl is really at the root of the phenomenon is proved by the fact that this precipitation is also hindered by non-acid substances, if these latter contain several hydroxyl groups, *e.g.* sugar, glycerine, etc.

CHAPTER IX

THE METALS OF THE IRON GROUP

1. *General Properties*

THE metals of the iron group form compounds with sulphur which are decomposed by dilute acids but not by water. They are therefore brought down by sulphide of ammonium, but not by sulphuretted hydrogen in an acid solution.

The laws which regulate the solubility of the metallic sulphides in dilute acids are the same as those which hold generally for the solubility of the salts of weak acids (p. 81), only here the conditions are simplified in that—sulphuretted hydrogen being a gas—the concentration of its solution cannot exceed a certain limit which is dependent upon the absorption-coefficient, at least so long as it is used at atmospheric pressure. If it is employed under higher pressure, zinc (for example) can be precipitated by it even from an acid solution; on the other hand, when the conditions are such that the sulphuretted hydrogen exerts only a certain very small pressure, the sulphides of lead and antimony (*e.g.*) become soluble in acids. The natural regulation of the concentration arising from sulphuretted hydrogen being a gas is one great reason of its value in analytical chemistry.

The solubility in acids of those sulphides of the metals which are insoluble in water depends upon the counter-ionising influence which these acids exert upon the sulphuretted hydrogen, and is therefore proportional to their strength or degree of ionisation. It thus increases with increasing concentration of the acids. Both of these statements can be conjoined in the one —that the solvent action is proportional to the concentration of the hydrogen ions in the solution. As the hydrogen ions become augmented the sulphur ions diminish, and solid sulphide must pass into solution in order that equilibrium may be re-established.

For the rest, the metals of this group form for the most part divalent ions of the magnesium type, but some of them also trivalent ions of the type of aluminium. The tendency to form complex ions is rather pronounced, and the result of this is a number of anomalous reactions. Cyanogen and ammonia in particular take part in the formation of such compounds. The precipitation of the hydroxide is prevented in every case by the presence of non-volatile organic acids. But these do not prevent the precipitation by ammonium sulphide, this being explained by the solubility of the sulphides being much smaller than that of the hydroxides, *i.e.* by the concentration of the metallic ions.

2. *Iron*

Iron forms an unusually large number of different ions. Apart from the fact that it can exist alone either as a divalent or trivalent positive ion, it shows a predilection for building up complex ions of various kinds, some of which are remarkably stable. Every iron compound can, however, be converted into ferrous sulphate, a salt of divalent iron, by heating with sulphuric acid; if at the same time oxidising agents are present, ferric sulphate is produced, which can be readily recognised by its characteristic reactions.

The ferrous ions follow those of magnesium in most of their reactions. They give an amorphous hydroxide, which passes very readily into the hydroxide of the trivalent iron, the colour being thereby changed from white through greenish-black to yellow-brown. They also form a sparingly soluble ferrous-ammonium phosphate, which is produced under similar conditions to the magnesium compound. Iron differs mainly from magnesium in being precipitable by sulphide of ammonium, which gives a greenish-black precipitate of hydrated ferrous sulphide, soluble in even very dilute acids. This therefore never comes down in acid solutions; even the "neutral" salts of iron possess as a rule a sufficiently acid reaction to prevent the production of the precipitate. In very dilute solutions the precipitate is obtained in colloidal form, into which the ordinary sulphide also soon passes upon washing; on account of this, and of the readiness with which it undergoes oxidation, it cannot be made use of for the ultimate quantitative separation of iron. For the latter purpose the iron is always first converted into the ferric state, and precipitated as the reddish-brown ferric hydroxide; the precipitation must be done in a warm solution, otherwise basic salts are formed, and it is incomplete. Potash and soda cannot be taken for this, for ferric hydroxide adsorbs these abundantly, and it is impossible to get rid of them entirely by washing. If, therefore, for any reason potash or soda has to be used in the first instance for precipitation, the precipitate must—after filtration—be redissolved in hydrochloric acid and thrown down again by ammonia. There being now only a very minute quantity of solid alkali present, the adsorption may be disregarded.

The ferric ion resembles that of aluminium in its reactions. Like the latter it is a very weak base, unable to form a carbonate in aqueous solution. The salts, even those of strong acids, are more or less hydrolysed in aqueous solution into free acid and colloidally dissolved ferric oxide; and this decomposition increases rapidly with rise of temperature until — when the acid is a weak one — the iron is completely separated. Should the acid be strong, the same result can readily be attained by adding sodium acetate to the solution. In this case it is especially important to filter hot, since a large part of the oxide would otherwise pass into solution upon cooling. The method is followed when for any reason it is inadmissible to make the solution alkaline.

Sulphuretted hydrogen reduces the ferric ion to the ferrous with separation of sulphur, which renders the liquid milky; sulphide of ammonium gives a precipitate of greenish-black hydrated ferrous sulphide, or—in very dilute solution—a black-green coloration.

Of the complex ions containing iron as a constituent, the compounds with cyanogen — ferrocyanogen and ferricyanogen—are of special importance. They are among the most stable of all compound ions. The amount of iron ion present in their solutions is less than in the aqueous solution of even the least soluble iron salt, so that all the compounds of iron dissolve in cyanide of potassium. It is true that this solution does not take place instantaneously (as we invariably find when dealing with a reaction which is not a purely ionic one), but still it goes on quickly enough to be applied in analysis. The solution in potassium cyanide does not show a single one of the ordinary reactions of iron, this being a necessary consequence of the condition first referred to.

We have thus the remarkable circumstance that iron can be tested for by means of a reagent which contains iron itself. The ferro- and ferricyanides of the heavy metals are sparingly soluble and usually of brilliant colour. With ferrous salts ferrocyanogen gives a white precipitate which rapidly becomes blue by oxidation, and with ferric salts a dark blue precipitate; ferricyanogen yields a blue precipitate with ferrous salts, but only a dark brown coloration with ferric, this last colour being due to the non-ionised portion of the ferric ferrocyanide produced. All those precipitates are amorphous and pass very readily into a colloidal mud, hence they cannot be washed; they are therefore only suited for qualitative and not for quantitative purposes.

Iron in the ferrous state can be estimated volumetrically with great ease and exactitude by means of potassium permanganate; should the iron in the solution be ferric, it must first be reduced, this being most conveniently done by zinc dust free from iron. The process depends upon the transformation of the ferrous into the ferric ion on the one hand, and the conversion of the permanganate into manganous salt on the other, thus—

$$2KMnO_4 + 10FeSO_4 + 8H_2SO_4 = K_2SO_4 + 2MnSO_4 + 5Fe_2(SO_4)_3 + 8H_2O.$$

One combining weight of permanganate is therefore equal in this case to five atoms of iron. The solution must be acid with sulphuric acid, but not with hydrochloric, since permanganate oxidises the latter in presence of iron salts. It is a catalytic reaction which goes on here, but very little is known yet of the laws which regulate it. The fact that oxalic acid can be titrated with permanganate in hydrochloric acid solution with perfect sharpness and without a trace of chlorine being set free, shows that the reaction abovementioned is not due to the hydrochloric acid and permanganate alone.

3. *Chromium*

The formation of complex ions is shown even more by chromium than by iron, for, besides the di- and trivalent cathion Cr, we have the divalent anion of chromic acid CrO_4, and the likewise divalent anion of dichromic acid Cr_2O_7. The two latter are to be looked upon as absolutely distinct compounds.

For the purposes of analysis it is unnecessary to consider the divalent chromium ion, for it changes so readily into the trivalent, that it can in fact only be obtained when special precautions are taken. The trivalent chromium ion resembles in its behaviour the other trivalent ions of aluminium and ferric iron, being somewhat weaker than the first and somewhat stronger than the second. Various bases of the general composition, $nCrO_3H_3 - mH_2O$ result from chromic oxide by condensation, and these are to be looked upon as the hydroxides of compound chromium-oxygen ions. That we have to do with new ions here, and not merely with basic salts, is shown by the change in colour and in analytical properties, and by the fact that the conversion of the one kind of salt into the other takes place gradually and not instantaneously. It is also of importance for practical analysis that chromic oxide combines very readily with polybasic acids to form complex acids, which show neither the reactions of chromium nor those of the acid in question. This takes place very easily with sulphuric acid, for example; when chrome alum is heated, the potassium salt of a chromi-sulphuric acid is formed, whose aqueous solution does not give the tests either for chromium or for sulphuric acid. Such compounds are readily decomposed by fusion with alkaline carbonate.

Chromic oxide dissolves in alkalies to a green solution, for the same reason that alumina dissolves in these. Precipitation takes place when this solution is boiled, because a less highly hydrated oxide is produced, in whose solution the chromium ions have a smaller concentration than in the alkaline liquid. The latter is, therefore, super-saturated with respect to this second oxide, which must consequently come down. Ammonia dissolves only traces of chromic oxide; the complex chrom-ammonium compounds, of which a great number are known, are formed in another way. Unlike potash and soda, ammonia is too weak a base to form a salt in this case.

The ion CrO_4 of chromic acid is yellow in colour, and the solubility of the chromates is somewhat the same as that of the sulphates. This ion is only stable in a neutral or basic solution; when it meets with hydrogen ions, two of these act upon two of the CrO_4 ions, with the formation of water and of the ion Cr_2O_7, which has a red colour. In consequence of this,

L

chromic acid is a weak acid, and the chromates which are sparingly soluble in water dissolve readily in acids. Barium chromate is unsuited to the separation of chromic acid, as it cannot be washed well; mercurous chromate is better, but must be washed with a solution of the nitrate, on account of its solubility. When it is ignited, chromic oxide is left behind.

Chromium forms complex ions with cyanogen similar to those of iron and cyanogen, but less stable.

4. *Manganese*

Unlike chromium, the divalent manganese ion is the more stable; the trivalent ion is so weak that its salts are incapable of existing in aqueous solution, being immediately decomposed by hydrolysis. There are only a few manganic salts which exist as well defined compounds,—the phosphate more particularly.

The manganous ion is pale pink in colour, and approximates more to magnesium in properties than to any other element; its behaviour towards ammonia in particular agrees almost perfectly with that of magnesium. The ammoniacal manganese solution becomes, however, muddy with a brownish precipitate on exposure to the air, from the separation of insoluble manganic hydroxide.

Manganous sulphide is the most soluble of all the metallic sulphides of this group, and therefore only separates with sufficient completeness in presence of excess of ammonium sulphide, and after prolonged standing; it must also be washed with water containing sulphide of ammonium, to prevent any going into solution. Further, its precipitation is prevented by very small quantities of acids, even comparatively weak ones.

Manganese forms with oxygen two different ions of the formula MnO_4, which have the same composition, and only differ in their valencies, the one being mono- and the other divalent. Notwithstanding their identity in composition they possess very different properties. The monovalent ion MnO_4 has an intense red colour, and resembles the ion of perchloric acid in its behaviour, while the divalent ion is as intensely green, and shows an analogy to the ion of sulphuric acid. Divalent MnO_4 is only stable in an alkaline solution; in an acid one it passes into the monovalent ion. Since half of the hydrogen ions must disappear in this latter case, the requisite oxygen is taken from another portion of the compound, which is thereby reduced to manganese peroxide.

The pronounced colours of the manganates and permanganates afford a convenient means for the recognition of any kind of manganese salt; the latter is converted into manganate upon fusion with sodium-potassium carbonate, the bright green colour of the fused mass showing the presence of manganese. Permanganic acid, is formed when manganese compounds are warmed with nitric acid and peroxide of lead, the liquid thereby becoming red. Chlorine compounds disturb this reaction, and must therefore be separated beforehand.

On account of its rapid oxidising action potassium permanganate is much used for the volumetric determination of oxidisable substances like iron, oxalic acid, etc. Iron is oxidised by it almost instantaneously, but oxalic acid requires a measurable time for the action to complete itself; towards the end of the titration the rate of reaction increases very distinctly. This arises from the accumulation of manganese salt (from the reduction of the permanganate), by which the action is hastened catalytically; the same quickening effect is noticeable if some manganous sulphate is added before the titration is begun. An excess of free acid likewise facilitates the process in proportion to the degree of concentration of the free hydrogen ions.

The strong colour of the permanganate itself renders the use of any indicator unnecessary; this is in fact one of the few cases of titration without an indicator.

The permanganate ion can be detected by the spectroscope with still more delicacy than by the naked eye. Its spectrum, which is shown equally by solutions of every salt of the acid, seeing that the same coloured ion is present in each, contains five dark bands in the yellow and green, and shows distinctly at a dilution when the colour is imperceptible by the eye alone.

5. *Cobalt and Nickel*

In cobalt and nickel we no longer find the capacity for forming trivalent ions. They do, it is true, form higher oxides; these are, however, not of a basic nature, but of the character of peroxides, being insoluble in dilute acids, and giving off chlorine with hydrochloric acid.

In these metals we have therefore to deal only with divalent ions, besides with some complex compounds possessing special properties of reaction. The cobalt ion is red, and that of nickel emerald green.

The non-ionised cobalt salts are for the most part dark blue; hence, as their solutions become concentrated, all the causes which go to diminish ionisation tend to change the colour from red to blue. Warming on the one hand, and the addition of strongly ionised salts with a like ion on the other, are among those causes. The action is best seen by adding concentrated hydrochloric acid to cobalt chloride. Although it is contrary to the generally prevailing view, there is no contradiction in ionisation being reduced by warming; this has in fact been often proved for the case of ionic dissociation.

Cobalt and nickel show a striking peculiarity in that, while their salts are not thrown down by sulphuretted hydrogen from an acid solution, the sulphides, after being once precipitated, are no longer soluble in dilute acid. This anomaly has still to be explained. It may however be surmised, on the one hand, that the sulphides undergo a change into a less soluble form immediately after precipitation, and on the other that they only exist in a sparingly soluble form, but that exceptionally persistent super-saturation phenomena in respect to the metallic sulphide in course of formation are produced in the acid solution. The latter of the two conjectures is the less probable, seeing that the sulphides are precipitated without any difficulty from an acetic acid solution. An investigation directed to the clearing up of this anomaly would be of some interest.

The capacity of forming complex ions is more developed in cobalt salts than in those of nickel, and upon this difference are based the methods for separating the otherwise very similar metals. The most convenient of these methods consists in treating the mixed solutions with nitrite of potassium in acetic acid solution. Potassium-cobalt nitrite is thus formed, or—more correctly—the potassium salt of a nitroso-cobaltic acid, a salt which is insoluble enough in presence of excess of potassic nitrite. This precipitation goes on but slowly, and the liquid must therefore be allowed to stand for several hours in order that the separation may be practically complete. This is a proof that we have to do here with the formation of a complex salt, and not with an ordinary ion reaction. Nickel gives no such insoluble compound under the circumstances.

Another method of separation is based upon the different behaviour of the complex cyanogen compounds. That of cobalt is extremely stable, and is not decomposed by acids even on boiling, while the corresponding nickel compound precipitates under those conditions the sparingly soluble nickelous cyanide.

The same difference in the stability of the complex ions shows itself in the ammonium compounds. Both metals are precipitated by ammonia from their solutions as hydroxide in the first instance, which redissolves in excess of the reagent. While, however, the nickel-ammonium compounds are so readily decomposable that in the solid state they lose ammonia in the air, those of cobalt become oxidised to stable complexes which are unaffected even on warming with alkali. The compounds just mentioned belong, too, to quite different types.

6. *Zinc*

Zinc forms only divalent ions; higher oxidation products are unknown in its case. Zinc hydroxide may give up hydrogen as an ion, whereby the very faintly acid ion ZnO_2'' is produced, and we also find zinc as a constituent of complex ions along with cyanogen, ammonia, etc.

Accordingly zinc oxide, which is insoluble in water, dissolves in potash and soda as well as in ammonia. The reason for this is, however, different in the two cases, the solubility being due in the first to the formation of the ion ZnO_2, and in the second to the production of complex zinc-ammonium ions. These latter are much more stable than the corresponding magnesium ones; hence the hydroxide is not split up hydrolytically, and zinc oxide dissolves in ammonia although no excess of ammonium salt is present.

Sulphide of zinc is less soluble than the other sulphides of this group. The greater part of the zinc separates out when sulphuretted hydrogen is passed through solutions of neutral zinc salts of the strong acids.

It depends upon the nature of the acid of the zinc salt how much zinc will still remain in solution. For the equilibrium of the solution with the solid zinc sulphide is determined by the product of the concentrations of the zinc and sulphur ions; and sulphuretted hydrogen being a very weak acid, the concentration of the sulphur ions is inversely proportional to the concentration of the free hydrogen ions of the acid which has been liberated. For the reasons given on p. 139, the concentration of the total amount of sulphuretted hydrogen may be considered as being constant. The less therefore the acid ionises and the more concentrated the solution of zinc salt is, the less zinc escapes precipitation. Further, since the ionisation of weak acids can be lowered at will by the addition of their neutral salts, the well-known old method of precipitating zinc in presence of an excess of sodium acetate is thus shown to be theoretically justified.

In order to completely prevent the precipitation of zinc salts by sulphuretted hydrogen, it is necessary to add a sufficient quantity of a strong acid. The free hydrogen ions of this then lower the ionisation of the hydrogen sulphide to such an extent that the value of the solubility-product is never reached, in spite of there being an abundance of zinc present. The amount of added acid must of course be approximately proportional to the amount of the zinc salt.

CHAPTER X

THE METALS OF THE COPPER GROUP

General Properties

THE metals of this group are distinguished from those of the iron one by the insolubility of their sulphides in dilute strong acids. From what has been already said, this difference is merely one of degree; the intermediate stages that we should expect are to be found in cadmium and lead. For the rest, the metals of this group differ very considerably among each other, so that it is hardly possible to give a general description of them.

Some of the metals of which we have to speak here belong to the so-called "noble" metals, and we might also refer to the others as being "nobler" than those of the iron group. This word, although indefinite, gives expression to a perfectly definite property of the metals, which may be termed their ionisation tendency, the measure of the latter being the free energy given out through the transition of an equivalent of the metal into the ionic state. The greater this is, the more easily and quickly will the metal pass into the ionic condition, and *vice versâ*. This tendency has its greatest value in the case of potassium; it is smaller with aluminium, zinc, tin and cadmium, almost zero with lead, and negative in the case of copper, antimony, bismuth, silver, gold, etc., *i.e.* free energy is given out when the ions of the latter metals change into the ordinary metallic state. The point at which the noble metals begin is not, however, determined by this, but by whether or no the metal is oxidised by free oxygen.

It is thus apparent that the metals with positive ionisation tendency are those which dissolve in acids with the evolution of hydrogen; this arises from the fact that the ionisation tendency of hydrogen is almost zero. In other respects the order of ionisation tendency is the same as that of the electro-chemical series of the metals, of which latter it is the exact expression.

Certain metals in particular solutions show deviations from the ordinary electro-chemical series. This always arises when the metal in question dissolves in the liquid to a complex compound, and the displacement is invariably in the direction of the metal comporting itself as if it were less noble. The reason of this is that the quantity of free energy referred to above is here inversely as the concentration of the ions in the liquid; it is greater the smaller this concentration becomes. If, therefore, there is present in the liquid a reagent which takes up the ions to the same extent as they are produced, an increased ionisation tendency remains permanent and the metal behaves like a less noble one. The converse of this cannot occur, for, although the concentration of the ions can be reduced to any degree, an insurmountable limit to their increase is very soon reached because of the limited solubility of the metallic salt; consequently there is no displacement of the position of the metal in the direction of the nobler ones.

The most striking phenomena of this kind are shown by cyanide of potassium. After what has been already stated, it is hardly necessary to say that this is due to the unusual facility with which cyanogen forms complex compounds with the metals.

1. *Cadmium*

Cadmium is very like zinc in its reactions, only its sulphide is less soluble, and therefore precipitates more completely from an acid solution than that of the latter metal. On the other hand, it requires a very considerable concentration of free acid to prevent the precipitation or to redissolve the precipitated cadmium sulphide. In other points the laws given for zinc sulphide hold good for the sulphide of cadmium.

In another respect we find here a phenomenon which is slightly indicated by zinc, more distinctly by cadmium, and which has a decisive influence upon the reactions of mercury, viz. the small ionisation of the halogen compounds. While no difference in this respect is noticeable between the oxygen and the halogen salts of the metals that have been already mentioned, we find a difference here, and it is necessary to bear this difference in mind if we would discuss the analytical behaviour of the metals properly.

The action which a slight ionisation of a soluble salt exerts is this: sparingly soluble precipitates are produced in its solution more imperfectly and with greater difficulty, and such precipitates are much more soluble in the reagents which go to build up the weakly ionised salt (*e.g.* in the corresponding acids), than would be the case under ordinary conditions. This difference is not yet very distinct in the case of cadmium; the chloride comports itself in almost exactly the same way as the other salts do, and only the iodide—which ionises least—shows measurable deviations. Hydriodic acid dissolves cadmium sulphide much more abundantly than either hydrochloric or nitric acid of equal concentration, and a solution of iodide of cadmium can only be precipitated slowly and incompletely by sulphuretted hydrogen. Hittorf pointed this out a considerable time ago, but it can be explained only by the dissociation theory.

Cadmium has not much tendency to form complex ions. The hydroxide is indeed soluble in ammonia, but the sparingly soluble salts like the carbonate are not so in any considerable degree. The complex cyanogen compound too, whose potassium salt has the composition $K_2Cd(CN)_4$, is less stable than many similar substances, *i.e.* the ion $Cd(CN)_4''$ is split up into $Cd^{..}$ and $4CN'$ to a sensible extent. For, in spite of the relatively great solubility of cadmium sulphide, this complex cyanide is decomposed by sulphuretted hydrogen with separation of the sulphide; cadmium ions are thus present in materially greater concentration in its solution than in the aqueous solution of cadmium sulphide.

2. *Copper*

Copper can form mono- and divalent ions—cuprous and cupric. The monovalent resemble those of silver and the monovalent mercury ions, while the divalent are like those of the magnesium group. The one ion readily changes into the other.

Of the salts of monovalent copper—the cuprous salts—only the halogen compounds are known, and these become less soluble with increasing weight of the halogen. The iodide is sufficiently insoluble to be made use of for the analytical separation of copper. When iodide of potassium is added to a cupric salt, the cupric and iodine ions react with one another to produce cuprous iodide and free iodine thus:

$$Cu^{\cdot\cdot} + 2I' \rightleftarrows CuI + I.$$

The reaction is imperfect, as the reverse process may go on at the same time; in order to make it complete, one of the products of the reaction must be removed. Sulphurous acid is therefore added, which converts the iodine into iodine ions (*i.e.* hydriodic acid); the concentration of one of the compounds on the left-hand side of the equation-formula is thus raised at the same time, and the separation of the sparingly soluble salt made more complete.

Advantage may be taken of the same reaction to separate the iodine from a mixture of halogen compounds by distilling the latter with an excess of sulphate of copper. The reaction is completed in this case by the mechanical removal of one of the two resulting products (the free iodine).

Similar processes go on when cupric ions meet with those of cyanogen or thiocyanogen. In the first case, as in that of iodine, half of the cyanogen is set free and escapes as gas. In the second, complex decomposition products result, unless care is taken to reconvert the thiocyanogen into the ionic state by adding some reducing agent.

The opposite process—the conversion of cuprous into cupric ions—takes place when cuprous oxide is treated with strong oxygen acids. We have thus the reaction:

$$2Cu^{\cdot} = Cu^{\cdot\cdot} + Cu,$$

i.e. two monovalent cuprous ions are transformed into one divalent cupric ion and metallic non-ionised copper. By following some suitable procedure (*e.g.* by the interaction of silver sulphate with cuprous chloride), it should be possible to obtain dissolved cuprous sulphate in very dilute solution, but so far as I am aware this has not yet been tried.

Copper also forms two sulphides corresponding to the two oxides, but the second is not obtained pure in aqueous solution, the precipitate consisting in part of cuprous sulphide mixed with sulphur. So when copper is estimated quantitatively as cupric sulphide, the latter must be ultimately reduced to the cuprous salt by ignition in a current of hydrogen.

Although cupric sulphide is far less soluble than sulphide of cadmium, its precipitation can nevertheless be prevented by hydrochloric acid of medium concentration. In such a case the sulphide can then be brought down by dilution with water, as this lowers the concentration of the hydrochloric acid, while that of the sulphuretted hydrogen remains the same—assuming that the gas continues to be passed through the liquid to saturation.

Both of the copper ions, mono- and divalent, form complex ions with ammonia; the first are colourless, but undergo oxidation very quickly into the second, which are blue. These compounds are so stable that most of the sparingly soluble copper salts dissolve in ammonia, the sulphide being an exception. Copper also enters with equal readiness into the hydroxyl of organic and inorganic hydroxy-compounds, from which

—excepting by sulphuretted hydrogen—it cannot be thrown down by the ordinary precipitants.

Of the complex compounds, the cyanogen-cuprous ion is dissociated to an exceptionally small extent into copper ions. Consequently all copper salts, including cuprous sulphide, are soluble in an excess of cyanide of potassium. This distinguishes copper from all the other metals of the group.

3. *Silver*

Silver forms—as such—only monovalent ions, but it has a great tendency to build up complex ions. It is characterised from an analytical point of view by the great insolubility of its halogen salts, which increases with the atomic weight of the halogen.

If we throw down a mixture of halogen compounds of different solubilities with (an insufficient quantity of) nitrate of silver, the precipitate will contain chiefly the halogen of higher atomic weight. But it is impossible to effect a complete separation of the latter in this way, for the halogen ions remain in the liquid in the ratio of the solubility-products of their silver compounds. If in a solution like sea water, for example, a small quantity of bromine is present along with a large quantity of chlorine, the former is precipitated very incompletely indeed by the addition of silver nitrate. In such a case it is first necessary to raise as far as possible the ratio of bromine to chlorine in the liquid under examination, *e.g.* by evaporating down and extracting the residue of dry salts with alcohol. If we could be certain that the precipitate was in the right chemical equilibrium with the liquid, the amount of bromine remaining in the solution could be calculated, seeing that it must then form a perfectly definite fraction of the chlorine present. It is assumed here that the bromine is present in not less quantity than corresponds to the equilibrium. Should this not be the case, the precipitate will consist of pure chloride of silver; and this can be applied to test whether the condition just mentioned has been fulfilled or not.

Silver combines with ammonia to a complex ion which contains four molecules of the latter to one atom of silver. This compound is among the more stable of its kind; in its solution the silver ions have a smaller concentration than in the aqueous solution of silver chloride, which follows from the solubility of the latter in ammonia. In the case of silver bromide the solubility is approximately such that the concentrations of the ions are equal. Iodide of silver is decidedly less soluble, and is therefore hardly dissolved by ammonia in a measurable degree.

The complex ion which silver forms with the anion of the thiosulphates is even somewhat more stable than the ammonia compound, the silver uniting directly with the sulphur. Hence sodium thiosulphate can dissolve not only all the silver compounds that are soluble in ammonia, but also some which are not soluble, or at least very sparingly so, *e.g.* silver bromide.

The most stable of the complex silver compounds is the silver cyanide ion, which has the composition $Ag(CN)_2''$. It is formed so quickly and easily that the reaction can be used for the titration of cyanogen. The liquid is made alkaline and a solution of silver nitrate run in; so long as the cyanogen ions are present in excess the solution remains clear, but as soon as the ratio of one atom of silver to two molecules of cyanogen becomes exceeded, silver cyanide is thrown down.

Cyanide of potassium dissolves all the salts of silver with the exception of the sulphide, the latter being—next to sulphide of mercury—the most insoluble sulphide of this group. The behaviour of silver salts with the thiosulphates has already shown us what a great affinity silver has for sulphur. For the same reason metallic silver decomposes sulphuretted hydrogen, with liberation of hydrogen, just as zinc breaks up hydrochloric acid; the character of the "noble" metals has entirely disappeared here. Silver must also dissolve in a solution of cyanide of potassium with evolution of hydrogen; experiments in this direction show in fact a distinct solubility.

Silver salts are always used to test for the halogens, although—as already stated—they only react when these are present in the state of ions. The slow precipitation that salts of silver give with organic halogen compounds (which cannot be regarded as salts in the ordinary sense), seems to me a proof that even such compounds can be dissociated in minute degree.

4. *Mercury*

Mercury forms mono- and divalent ions, the former resembling the ions of silver, and the latter those of cadmium. There are also many similarities between mercury and copper. A special characteristic of mercury is its tendency to form compounds of low ionisation, which gives rise to a great number of "abnormal" reactions.

The monovalent or mercurous ion yields like that of silver sparingly soluble halogen compounds, whose solubilities follow the same order as those of the silver series. They differ from the corresponding silver salts by the formation of black insoluble ammonia compounds; while silver chloride dissolves in aqueous ammonia, mercurous chloride is turned black by it. There are certain resemblances to copper in respect to the mutual transformations of the mono- and divalent ions, but at the same time some striking differences. Thus, while mercurous oxide passes readily into mercuric, the mercuric halides—in contradistinction to the copper compounds—are far more stable than the mercurous. Mercurous sulphide has no existence, breaking up as it does at the moment of formation into mercuric sulphide and metallic mercury. Mercuric sulphide, on the other hand, is an exceptionally stable compound; it is the only sulphide which does not dissolve in nitric acid. It resembles to some extent the sulphides of the next group in being dissolved by sulphide of potassium (although only in concentrated solution) in presence of caustic alkali; it is reprecipitated again on diluting. Sulphide of ammonium does not dissolve it.

The divalent mercuric ion shows itself in the oxygen salts—which are dissociated normally—to be a very weakly basic ion; its salts are split up to a great extent hydrolytically in aqueous solution, for a clear solution can only be obtained by having an excess of free acid present. The halogen compounds, on the other hand, are perfectly stable on dissolving, although in their reactions they deviate in many points from the oxygen salts. The electric conductivity shows that the mercury halides are ionised in an extremely slight degree, so that their solutions only contain mercury ions in very small concentration. We have already seen that compounds of slight ionisation always result when their constituent ions meet, and so the oxygen salts of mercury assume the reactions of the halides whenever they are brought into contact with soluble halogen compounds. A considerable amount of heat is given out here, in contradistinction to the ordinary behaviour of neutral salts upon interaction, when the law of thermal neutrality holds good, and no heating effect at all is produced.

Mercuric oxide acts as a very weak base with respect to most acids; when, however, it is added to soluble halogen compounds, the liquid becomes at once strongly alkaline. This action is weakest with chlorides and strongest with iodides, iodide of potassium being decomposed by mercuric oxide to the extent of 90 per cent. The reaction depends, on the one hand, upon the slight ionisation of the corresponding mercury compounds, and, on the other, upon the union of the latter with excess of the alkaline halide to form the alkali salts of very stable mercuric-halogen hydracids. The stability of these complex compounds likewise increases with increasing atomic weight.

The same cause is at the root of the reverse phenomenon—that the mercuric halides are only decomposed by alkalies with difficulty. Mercuric chloride requires a considerable excess of alkali for this, while the iodide is not attacked at all. It may, however, be decomposed either by sulphuretted hydrogen or by alkaline sulphides.

Lastly, the same considerations explain the reactions on which Liebig's process for the volumetric determination of chlorine (as ion) are based. A solution of mercuric nitrate in a certain excess of nitric acid gives a precipitate with urea, while none ensues between urea and mercuric chloride. The reason of this is that more mercury ions are contained in the first solution, which is split up hydrolytically, than correspond to the solubility-product of the sparingly soluble urea compound; while in the solution of mercuric chloride the mercuric ions are present in very small number only, and the critical value is never reached. If, therefore, mercuric nitrate is added to a solution containing a chloride together with urea, no precipitation takes place so long as there are still chlorine ions present to form non-ionised mercuric chloride, but after that the slightest excess of nitrate produces a precipitate.

The great affinity of mercury for sulphur causes mercuric oxide to react with sodium thiosulphate and sodium sulphite in the same way that it does with potassium iodide; *i.e.* a strongly alkaline liquid results. And a similar action takes place with potassic cyanate, thiocyanate and nitrite; in all these cases complex compounds are formed whose dissociation into mercury ions is but very small. In organic compounds, too, which contain hydrogen combined with nitrogen or sulphur, mercury replaces the hydrogen with great readiness; from the solutions of the compounds thus formed mercuric oxide is not precipitated by alkali, or at least only incompletely.

The action of iodide of potassium upon mercurous salts depends upon similar causes, half of the mercury being separated here in the metallic state. The reaction is like that of acids upon cuprous salts; two monovalent mercurous ions give one atom of metallic mercury and one divalent mercuric ion, the latter passing immediately into potassium-mercury iodide.

When mercuric salts are thrown down by sulphuretted hydrogen, we get at first a white precipitate, which becomes by degrees red, brown, and finally black. The white substance is a compound of mercuric sulphide with the other mercuric salt still present, the latter becoming gradually decomposed by the excess of sulphuretted hydrogen. Sulphide of mercury is not oxidised by the air, unlike almost all the other metallic sulphides, because it is more stable than either the oxide or the sulphate,—a point which follows from what has been already said.

In the cyanide mercury forms a complex compound of the most stable kind. The electrical conductivity of mercuric cyanide is so small as not to be measurable; and the solution of the salt is not precipitated either by caustic alkali or any other reagent, with the exception of sulphuretted hydrogen or alkaline sulphides. With potassic cyanide it forms the very stable salt, potassium mercuricyanide. It may be looked upon as the type of compound which is incapable of reaction for want of electrolytic dissociation.

5. *Lead*

Unlike mercury, lead has very little inclination to form complex compounds; its reactions are therefore almost all normal.

There is only the one kind of lead ion, the divalent; the higher oxide of lead is incapable of electrolytic dissociation. Like the hydroxide of zinc, lead hydroxide can, however, give up hydrogen and thus form an oxygenated anion—as follows from its solubility in alkalies. But lead does not possess the capacity—so general among the heavy metals—of forming complex stable compounds with cyanogen and ammonia. The only abnormal reaction that we have to consider is the substitution of hydroxylic hydrogen in organic hydroxy-compounds. The result of this is the formation of salts of complex acids containing lead which are soluble in alkaline liquids; the sparingly soluble lead salts dissolve, for example, in basic tartrate of ammonium. Sodium thiosulphate possesses a like solvent power for these, becoming converted by contact with lead salts into the salt of a plumbo-thiosulphuric acid, but this compound decomposes very quickly with the separation of sulphide of lead, so it is not made use of in analysis.

The sulphate serves for the analytical separation of lead. It has about the same solubility as sulphate of strontium, and one must consequently use a large excess of sulphuric acid for the precipitation, the acid liquid being ultimately displaced from the filter by alcohol. Lead sulphate resembles sulphate of barium in appearance, but is easily distinguished from the latter by its solubility in ammonium tartrate.

Lead sulphide is not one of the most insoluble sulphides, its precipitation being prevented by hydrochloric acid of moderate concentration. It must therefore be thrown down in dilute solution.

The halogen compounds of lead are not insoluble enough to be of much use for quantitative analysis. The iodide forms a soluble complex salt with iodide of potassium in concentrated solution, which is broken up again into its constituent salts by excess of water. Hence in a solution of potassic iodide of increasing concentration, the solubility of iodide of lead at first diminishes on account of the increase in iodine ions, and then augments through the formation of the complex salt.

6. Bismuth

The type of compounds which it forms places bismuth alongside of arsenic and antimony, which belong to the next group of metals. But the general law—that with increase of atomic weight the acid properties diminish—is exemplified in bismuth to such an extent that its sulphide can no longer form soluble thio-salts with the sulphides of the alkalies. It must therefore be reckoned for analytical purposes as belonging to the copper group.

Bismuth forms a trivalent cathion of very weakly basic character. Its salts are all strongly decomposed by water hydrolytically, with the production of sparingly soluble basic salts, this reaction being a characteristic of bismuth. In many cases the resulting compounds are to be regarded as salts of the monovalent ion $BiO^{\bullet}$, which shows certain points of resemblance with the ion of silver or of monovalent mercury. The chloride, $BiOCl$, in particular, not only resembles chloride of silver and calomel in its insolubility, but also to some extent in its appearance.

There is practically no tendency on the part of bismuth to form complex salts, neither cyanogen nor ammonia exerting any solvent action upon its sparingly soluble salts. Bismuth would be the only heavy metal which showed no abnormal reactions, were it not that the precipitation of oxide is prevented by organic hydroxy-compounds in its case also.

CHAPTER XI

THE METALS OF THE TIN GROUP

1. *General Properties*

THE metals of this—the last—group form, like those of the preceding one, sulphides which are insoluble in dilute strong acids, but which differ from the others by dissolving in alkaline sulphides. This solubility depends upon the production of thio-salts, *i.e.* salts of similar composition to the oxygen ones, but containing sulphur in place of oxygen. The alkaline salts thus obtained are soluble in water, and break up on addition of acid into metallic sulphide, which is thrown down, and sulphuretted hydrogen. The free thio-acid is really formed in the first instance, but it is unstable and immediately breaks up as stated. It may be asked why it should do this seeing that the neutral salt and free acid both contain the same ion, whose stability should not be affected by the mere presence of the other ion. The answer is that the other possible decomposition products—besides the metallic sulphides—viz. alkaline sulphide on the one hand or sulphuretted hydrogen on the other, are dissociated very differently; the last-named, being a very slightly ionised compound, is produced in the greatest possible amount, with the result that the complex is broken down. An excess of strong acid thus exerts a decomposing action for two reasons—(1) it reduces the ionisation of the sulphuretted hydrogen still further and so increases the tendency to decomposition, and (2) the rate of decomposition is augmented catalytically by the presence of hydrogen ions. Lastly, acids also check the inclination of the sulphides to pass into the colloidal state. We must remember, however, that too large an excess of acid has to be avoided, since some of the sulphides in question are soluble in strong acids.

The capacity of forming thio-salts is closely conjoined with the property of the same metal to give mainly oxides of acid character with oxygen. Just as these dissolve in alkalies, so do the sulphides dissolve in alkaline sulphides.

2. Tin

Tin forms divalent ions; its higher oxygen compound is an acid anhydride, but the existence of tetravalent tin ions is not altogether impossible. The properties of the stannous ion are peculiar to itself, and they bear no resemblance to those of the ions of the other metals already spoken of. The readiness with which stannous compounds pass into stannic is very characteristic, hence they act as powerful reducing agents.

Stannous hydroxide is soluble in alkalies, and it can therefore form an anion SnO_2''. The alkaline solution possesses extremely strong deoxidising powers, reducing even bismuth salts from their solutions with the production of a characteristic precipitate of black colour.

From the concentrated solution tin gradually separates out, a stannate being formed at the same time. The reaction may be regarded as the conversion of the divalent ion SnO_2'' of the stannites into the equally divalent ion SnO_3'' of the stannates, the requisite oxygen being abstracted from a second SnO_2 ion. The chemical equation is—

$$2K_2SnO_2 + H_2O = K_2SnO_3 + 2KOH + Sn.$$

The solubility of stannous hydroxide in alkalies will have already led to the deduction that it is a very weak base, *i.e.* a substance which yields up its hydroxyl with great difficulty. Hence the stannous salts have an acid reaction and are decomposed hydrolytically.

Sulphuretted hydrogen gives with stannous salts a dark brown precipitate of stannous sulphide, which is not soluble of itself in sulphide of ammonium; from yellow ammonium sulphide, however, it takes up sulphur, and then dissolves with the formation of ammonium thio-stannate. Since this is not a simple ion reaction, it requires a measurable time for completion, and so the mixture has to be gently warmed and the precipitate extracted more than once with fresh sulphide of ammonium. Acids throw down yellow tin disulphide from the solution so obtained.

Stannic acid exists in several modifications, which change readily the one into the other. It is scarcely at all soluble in water, properly speaking, but it very easily forms colloidal solutions from which it can be separated by the usual means, sulphuric acid and sulphates being the most efficacious for this purpose. The solution of tin tetrachloride in water contains, indeed, no measurable quantity of tetravalent Sn-cathions, as the thermo-chemical researches of Thomsen and the electrolytic ones of Hittorf have shown; but several of its reactions, more especially the reducing actions of solutions of stannous chloride, point to the presence of at least a small number of such ions. The solution must then at the same time contain a certain amount of non-ionised tin tetrachloride.

When a solution of the tetrachloride is neutralised by potash or soda, or ammonia, gelatinous stannic acid comes down. Since, as already stated, there is no considerable amount of ionised or non-ionised stannic chloride in the solution, this reaction may perhaps be explained by the circumstance that some colloids can remain dissolved in an acid solution for a very long time, while they coagulate quickly when the solution is made exactly neutral. Silicic acid behaves in this way; when a dilute solution of water glass is just neutralised, the liquid very soon coagulates, while it remains clear in presence of a large excess of acid. An explanation is, however, still wanting of the fact that the stannic acid readily redissolves in acids after having been precipitated by neutralising its solution. In alkalies (potash and soda) the precipitated stannic acid is of course easily soluble; ammonia is too weak for this, or, what amounts to the same thing, the stannic acid is too weak for ammonia.

3. *Antimony*

Antimony forms trivalent ions of very weakly basic character. In addition, it gives a pentoxide, which is the anhydride of a likewise very weak acid. Like phosphoric acid the latter exists in various modifications, which, however, change much more readily one into the other than the phosphoric acids do.

The salts of trivalent antimony are split up hydrolytically to such an extent by water that it requires a large excess of free acid to retain them in solution. In order to have them in a convenient form, use is made of the marked capacity which antimony possesses of substituting the hydroxylic hydrogen of organic compounds by the monovalent antimonyl radicle, SbO. The compounds thus obtained are so stable that they continue permanent even in acid solution, this being again a consequence of the very weakly basic character of antimony trioxide. Tartaric acid is generally used in analysis for this purpose; the resulting antimonyltartaric acid is undecomposed by water and dilute acids so long as an excess of tartaric acid is present, and hence solutions of antimony salts to which a sufficiency of tartaric acid has been added are no longer precipitated by water.

Sulphuretted hydrogen throws down the orange trisulphide from solutions of antimony salts. Since this is somewhat soluble in concentrated hydrochloric acid, the precipitation must be done in dilute solution. The trisulphide takes up sulphur from a solution of yellow sulphide of ammonium, and dissolves with the formation of thio-antimoniate; from this solution acids precipitate the yellow-red antimony pentasulphide, with evolution of sulphuretted hydrogen. The theory of these reactions has been given already.

If antimony trisulphide is dissolved in concentrated hydrochloric acid to saturation, and the solution then diluted with water, a precipitate of trisulphide comes down. Now, seeing that the hydrochloric acid and sulphuretted hydrogen, which are in equilibrium with the antimony, bear the same relation to one another in the diluted as in the concentrated solution, there appears at first sight to be no ground for this precipitation. The cause lies in the fact that the ionisation of the sulphuretted hydrogen, which is a very weak acid, increases much more rapidly with dilution than that of the hydrochloric acid does; the latter is already ionised to a great extent in strong solution, and its ionisation cannot therefore go much further. This is, however, only one of the factors in the question; a complete explanation would lead to considerations of a more abstract nature, which would be out of place in a book of this size.

Trioxide of antimony dissolves in alkalies; it is thus capable of forming an anion. The composition of the crystalline sodium salt shows the latter to be monovalent, and to have the formula SbO'_2. The solution acts as a reducing agent, the SbO'_2 changing into SbO'_3.

Of the salts of antimonic acid, the sodium one concerns us most. It is an acid salt of the type of the pyrophosphates, and is made use of as a test for sodium on account of its insolubility.

Antimony trifluoride is not precipitated by water. The investigation of the electric conductivity of its solution has shown this to be very small. The trifluoride is therefore dissociable only to a minute extent, and its solution contains too few ions to give with the hydroxyl ions of the water the solubility-product value for antimony oxide. It is still more stable in presence of excess of hydrofluoric acid, forming as it does a hydrofluo-antimonic acid (analogous to hydrofluo-boric acid), which yields even fewer antimony ions.

4. *Arsenic*

Arsenic stands midway between the metals and the non-metals; it hardly exists as an elementary cathion, but on the other hand it forms compound anions of various kinds. Those of the latter which are of importance in analysis are the ions of arsenious and arsenic acids and the corresponding sulphur compounds.

Arsenic trioxide dissolves more readily and abundantly in hydrochloric acid than in pure water. It therefore follows that a reaction takes place between the ions of the hydrochloric acid and those of the trioxide, even although the small electrical conductivity shows that but few of the latter can be present in the solution. The phenomenon cannot be explained by the lessening of the ionisation of the arsenious acid through the hydrochloric acid, for even if this relative alteration is a large one, the absolute increase of the non-ionised portion is almost nothing, because of the extremely small quantity of acid ions. And since there is equilibrium of solution equally with this portion and with that which is ionised, the solubility cannot alter appreciably from this cause. There thus remains only the one assumption—that arsenic trichloride is present in the solution both in the ionised and non-ionised condition. The fact that arsenic volatilises when a solution of the trichloride is heated, also tells in favour of this view. At all events elementary arsenic cathions are to be assumed as present in the solution in question, although there is as yet no known method for determining their concentration. The subject would be advanced by an investigation of the solubility of arsenic trioxide in other acids.

Sulphuretted hydrogen throws down the trisulphide from acid, but not from alkaline solutions. In neutral solutions—especially when there is not much foreign matter present—the colloidal sulphide is formed, and this passes through a filter; it coagulates, however, when acid is added. The trisulphide is very difficultly soluble in acids; hydrochloric acid of the concentration that dissolves antimony sulphide readily is without effect upon it—a fact that can be made use of in separating the two compounds. This behaviour depends not merely upon the sparing solubility of the sulphide of arsenic, but also to at least an equal extent upon the circumstance that arsenic forms cathions with greater difficulty than antimony.

Trisulphide of arsenic is soluble not only in sulphide of ammonium, but also in ammonia, and even in ammonium carbonate, this last reaction being utilised to separate it from the sulphide of antimony. The reaction is due to the circumstance that the oxygen in both arsenious and arsenic acids can be replaced by sulphur in almost any proportion without the conditions of solubility and stability undergoing any material change.

Arsenic acid is not at first precipitated by sulphuretted hydrogen, but a reaction gradually sets in, sulphur being separated and the trisulphide formed. This process is hastened by the presence of free acid and by warming, but is so slow at best that it is more convenient and efficacious to first reduce the arsenic acid to arsenious by some suitable reducing agent such as sulphurous acid.

No arsenic volatilises when a solution of arsenic acid is warmed, for there is no pentachloride, and arsenic acid is not volatile itself.

CHAPTER XII

THE NON-METALS

1. *General Properties*

THE capacity to form elementary cathions may be regarded as a characteristic property of the metals. On the other hand, elementary anions come solely from the non-metals. But just as the metals can form compound anions, so can the non-metals give rise to compound cathions. True, the number of these latter is much the smaller of the two; among inorganic compounds we have only ammonia, and among organic the substitution products of ammonia, together with the analogous organic compounds of phosphorus, arsenic and antimony, and the bases of the sulphur group. The iodonium bases, recently discovered by Victor Meyer, have also to be included here.

On account of their more complex nature, it is much less easy to classify the anions than the cathions, most of which are elementary. The most convenient arrangement is according to the valency, which keeps together the substances constituting natural groups, even if it also includes some that are less nearly related. We shall accordingly take first the monovalent halogens, whose compound anions are also mainly monovalent; second, the divalent sulphur group with the likewise divalent compound anions; third, the trivalent compound anions of the phosphorus group (trivalent elementary anions are unknown); and, last, the tetra- and polyvalent anions.

2. *The Halogens*

Chlorine, bromine and iodine constitute a closely connected group of monovalent anions, whose properties alter regularly with rise or fall of atomic weight. The specific reagent for them is silver, which gives sparingly soluble white to pale yellow precipitates that blacken on exposure to light, especially when there is an excess of silver present. The same reaction is shown by the mercurous, the thallous, and—in a lesser degree—by the cuprous ion. The monovalent bismuthyl, BiO, is also to be included here.

The tendency of the halogens to ionisation decreases with increasing atomic weight. Since, however, the solubility of the iodine compounds is usually least, it often happens that they appear to be more stable under certain conditions than the corresponding chlorine and bromine ones. With regard to this the rule holds that, in reactions in which free halogen is liberated, iodine is the weakest; in pure ion-reactions, on the contrary, *i.c.* in cases of double decomposition, iodine often holds the upper hand. For this reason iodide of potassium yields chloride of potassium and free iodine when treated with chlorine, while chloride of silver is converted into the iodide on digestion with iodide of potassium; chloride of potassium has no effect upon silver iodide.

The two conditions just mentioned therefore form the basis of the various processes for separating the halogens from one another. There are a number of weak oxidising agents, like the ferric and cupric salts, whose action is sufficient to overcome the slight tendency of iodine to ionisation, but which are unable to change bromine or chlorine ions into the free element. The action is usually only a partial one; it can, however, be made complete in practice by removing the free iodine as it is formed, and thus doing away with its mass-action. This is generally effected by distillation, but it may also be done by shaking up with another solvent, such as bisulphide of carbon.

The same method might be applied for separating bromine from chlorine if there was a suitable oxidising agent for the purpose, but no investigations have been made in this direction so far as I am aware. From Bancroft's measurements of the electromotive force of different oxidising and reducing agents, only iodic acid (*i.e.* potassic iodate and sulphuric acid) among the substances that he examined would be available. Possibly a solution containing hydrate of peroxide of manganese and sulphuric acid at the ordinary temperature might also do, the bromine being removed by shaking up with chloroform or some similar solvent.

For the quantitative determination of two halogens occurring together, indirect analysis may be resorted to with advantage, if there be not too great a difference between the amounts of the halogens present. The simplest form of this process is first to determine volumetrically the quantity of silver required for complete precipitation and then to weigh the precipitate. From the first measurement we can calculate how much the precipitate would weigh if it contained only one or the other of the halogens; the differences of these two numbers from that observed stand in the same ratio as the amounts of the two halogens. Or, the halogens may be completely thrown down by silver solution and the precipitate weighed; the latter then converted entirely into the salt of the stronger halogen by heating it in a stream of the latter and the precipitate weighed again. The calculation is similar to the one above.

An explanation has already been given of the behaviour of the silver compounds when one of the halogens is present only in very small quantity.

Fluorine differs greatly from the three halogens already spoken of. It does not form insoluble compounds either with silver or with the other metals that have been mentioned, but it does yield such with the metals of the alkaline earths, with which chlorine, bromine and iodine form soluble salts. Here, again, we have the same deviating behaviour on the part of the element with lowest atomic weight that was noticeable in lithium and beryllium. The usual test for fluorine depends upon the formation of the volatile silicon tetrafluoride, which decomposes with water into silicic and hydro-silicofluoric acids.

Of the free halogens iodine is the most easily tested for, because of the blue coloration that it gives with starch solution. The colour is that of an easily dissociable addition-product of the two substances; when warmed, the compound breaks up into its two constituents and the colour vanishes, to reappear again when the solution is cool. Iodide of starch apparently dissociates to a very considerable extent at the ordinary temperature also, for the iodine behaves in this compound almost like free iodine; in some processes which only complete themselves slowly, however, the retarding influence of the starch shows that the concentration of the free iodine has been diminished by the presence of the former.

Another very delicate reaction shown by iodine is its intense reddish-violet colour when in solution in carbon disulphide or chloroform, etc. Being much more soluble in these than in water, it is taken up by them from the latter almost completely when the two liquids are shaken up together; the division-ratio with carbon disulphide is 1 : 410. On the other hand, the ions of iodine are far more soluble in water than in any other solvent. Use may be made of this for the determination of very small quantities of iodine. The iodine is set free by a suitable oxidising agent and taken up with carbon disulphide, and this solution—after separation from the other—is titrated with thiosulphate until it becomes colourless, the mixed liquids being shaken up together after each addition of thiosulphate. Nitrous acid is used as the oxidising agent, which must not be allowed to remain mixed with the disulphide of carbon. The passage of iodine from one solvent to another is beautifully shown by gradually adding chlorine water to a dilute aqueous solution of an iodide, a little carbon disulphide having previously been dropped in. Free iodine is liberated at first and the disulphide becomes violet, but as more chlorine is added the solution ultimately turns colourless again, from the iodine changing into the ion of iodic acid.

Bromine does not affect the conversion of iodine into iodic acid so well as chlorine does, its tendency to ionisation being distinctly less; bromides of iodine can therefore exist in the solution without passing into hydrobromic and iodic acids, *i.e.* into the ions of bromine, iodic acid, and hydrogen. The result of this —especially in concentrated solutions—is that the carbon disulphide becomes yellowish-brown from dissolved bromide of iodine, which is not decomposed by water; and it is only after large dilution (when ionisation is promoted in a corresponding degree) that the reaction shown by chlorine takes place.

Free bromine is easily recognised by its odour and by the yellowish-red colour of its solutions. Like iodine it is much more soluble in carbon disulphide and similar solvents than in water, and can therefore be concentrated by shaking up with these, its recognition being thereby much facilitated. Its quantitative estimation is always an indirect one, the bromine being replaced by iodine through the addition of an iodide, and the iodine then titrated with thiosulphate. Free bromine cannot be titrated with the latter, because it does not convert it into tetrathionate, but into sulphuric acid, free sulphur, etc.

Iodine and bromine, and also chlorine in a lesser degree, are much more soluble in solutions of their salts and hydrogen acids than in water alone. This is a proof that part of the halogen is not present in the solution in its ordinary state; the portion remaining over and above that of the amount soluble in pure water must be there in some other form. It has been shown that the free halogen unites with the ion of the same name to a compound ion I_3 or Br_3, which is partially dissociated. The reactions of the free halogens, as we know them in aqueous solution, are therefore essentially reactions of these double compounds, although we must not forget that the latter split off free halogen very readily.

Chlorine in the free state is also recognisable at

once by its odour, and the quantitative estimation is effected in the same way as that of bromine, *i.e.* by determining the equivalent amount of iodine which it expels from iodide of potassium solution. On account of its volatility it is often condensed first in cold dilute alkali, with which it forms a mixture of (non-volatile) hypochlorite and chloride, this mixture giving up its chlorine again on treatment with acid. If, however, the liquid is allowed to stand for some time, a part of the hypochlorite changes into chlorate, which is only slowly decomposed by acids; hence in such cases it is very easy to get too low results.

3. *Cyanogen and Thiocyanogen*

The two compound ions cyanogen, CN, and thiocyanogen, CNS, resemble the halogens closely in many of their reactions, more especially in yielding silver salts just like those of the halogens.

Cyanogen is remarkable for the readiness with which it forms complex ions, in which the ordinary cyanogen reactions are no longer apparent. Thus we do not find in yellow prussiate of potash the poisonous properties characteristic of every compound containing cyanogen ions. The more important of these complex ions have been already described under the metals, where attention has also been drawn to their very different relative stabilities. Most of the analytical reactions of cyanogen depend upon the formation of such complexes.

One of the most convenient and delicate tests for a cyanide consists in the formation of ferric ferrocyanide or Prussian blue. To the liquid under examination an excess of ferrous and ferric salts and then caustic potash are added, the mixture being then warmed for a little. If cyanogen is present, potassic ferrocyanide is thus formed, and the characteristic blue precipitate appears after acidifying; when only traces of cyanogen are being dealt with, this blue precipitate is replaced by a bluish-green coloration. It is to be noted that the above alkaline solution must be warmed for some time, for the production of ferrocyanide is not a simple ion reaction, and therefore does not take place immediately.

Another extremely delicate test is to evaporate a little of the solution at a gentle heat along with an excess of yellow sulphide of ammonium. The cyanogen is thus changed into thiocyanogen, which is readily detected by its well-known reaction with ferric salts.

For the quantitative estimation of cyanogen we have either to precipitate with a silver solution and weigh the dried silver cyanide, or to follow the volumetric method given on p. 160.

Thiocyanogen is characterised by the deep blood-red coloration which it gives with ferric salts. This colour is due to the non-ionised portion of the salt, and is therefore weakened or intensified by any causes which go to increase or diminish the ionisation. Thus the red colour is lessened by adding a neutral salt like sodic sulphate to the liquid. For, the effect of the added sulphuric acid ions is to convert a portion of the ferric ions into non-ionised salt, since ferric sulphate—as the salt of a dibasic acid—is less dissociated than the thiocyanate. On the other hand, the reaction becomes more distinct when the liquid is shaken up with ether, for then the non-ionised red ferric thiocyanate is taken up by the ether, and new salt must therefore be formed in the aqueous solution.

When thiocyanate of potassium and a ferric salt are mixed in equivalent quantities, we by no means get the maximum colour effect; it becomes greater on adding an excess of either the one or the other, because the increase of one of the two ions causes a change in the equilibrium in the direction of an increased production of non-ionised ferric salt. No coloration is obtained at all with solutions of colloidal ferric oxide, since these contain no ferric ions, and the same applies to a solution of red prussiate of potash.

Thiocyanogen is determined quantitatively by precipitation with nitrate of silver, or—in presence of other substances precipitable by silver—by oxidation to sulphuric acid, the latter being then estimated in the usual way.

4. *The Monobasic Oxygen Acids*

The acids HNO_3, $HClO_3$, $HClO_4$, $HBrO_3$ and HIO_3 resemble one another just as the halogen acids do. Their chief characteristic lies in their forming almost only soluble salts; iodic acid, which stands at the outside limit, constitutes an exception, some of its salts, more especially the barium one, being sparingly soluble. Barium bromate is more soluble, and the chlorate the most soluble of the three.

The analytical reactions of these anions do not depend upon ion-reactions proper, but upon the readiness with which oxygen is given off, and the consequent production of substances that are easy to recognise. The salts of the ions ClO, ClO_2, ClO_3, ClO_4, BrO_3 and IO_3 pass, on heating, into salts of the halogens themselves, which can then be tested for in the ordinary way. It is noticeable here that the oxygen compounds are more stable the more oxygen they contain, this being just the reverse of what we might have expected, judging from the analogies in other groups.

The most convenient test for nitric acid is that with a ferrous salt in solution in concentrated sulphuric acid; when the liquid containing nitrate is poured carefully on to the top of this, a brownish-violet ring is produced where the two layers meet. The colour is due to the formation of a complex cathion which contains the elements of nitric oxide in addition to iron. This follows from the fact that all ferrous salts give the reaction, whatever their acid may be. The complex iron-nitric oxide ion is not very stable, being destroyed when the liquid is boiled. This arises from the small portion of nitric oxide present through ionisation being carried away by the steam; hence fresh nitric oxide must be set free in order that equilibrium may be re-established, and so on until the compound is entirely broken up. The same thing must take place when an indifferent gas is passed through the solution (although I have never heard of the point being actually investigated).

The quantitative determination of nitric acid is based upon the same reaction, either the amount of ferrous salt oxidised being estimated, or the evolved nitric oxide measured. The former method is the more convenient, but can only be followed in the absence of other oxidising or reducing substances; the latter—Schloesing's method—is more complicated but of wider application. The iodometric method may also be employed in certain cases.

The various oxygen compounds of chlorine are distinguished qualitatively by their different stability. Hýpochlorous acid is decomposed even by cold dilute hydrochloric acid with evolution of chlorine, chloric acid only upon warming (with hydrochloric), and perchloric acid not at all in this way. The quantitative estimation is made by measuring the oxidising effect, this being most easily done with hydriodic acid, *i.e.* potassic iodide and hydrochloric acid. Hypochlorous acid acts instantaneously, while chloric acid requires as an oxidising agent a considerable time.

Perchloric acid cannot of course be estimated in this way, but it may be thrown down as the sparingly soluble potassium salt by adding acetate of potassium and alcohol. It is necessary here to add a large excess of the acetate, perchlorate of potassium being comparatively soluble (*i.e.* from a quantitative point of view). If this process is objected to, the perchlorate has to be converted into chloride by heating.

Bromic acid is not decomposed very quickly by hydriodic acid, but both iodic and periodic acids break up with the latter instantaneously. In this reaction the bromic changes into hydrobromic acid, *i.e.* the bromine assumes the ionic state, while iodic and periodic acids allow their iodine to become free. The amount of iodide liberated is the same for bromic as for iodic acid, *i.e.* six atoms of iodine to a molecule of acid.

It is very noteworthy that the lower oxygen acids of chlorine and bromine are extremely weak acids; the addition of oxygen to the very strong hydrogen acids has thus had the effect of reducing the capacity for ionisation in an extraordinary degree. Nothing is known as to the cause of this phenomenon, which stands in striking contrast to the well-known acidifying action of oxygen, but it might be sought for in a change of valency in the halogen; the negatively acting sulphur of the alkyl sulphides, for instance, acquires a markedly basic character after being transformed into the tetravalent sulphur of the sulphines. The sudden transition only takes place, as a matter of fact, in the change from hydrogen acid to the lowest oxygen one; in the series of oxygen acids themselves the strength increases regularly with increase of oxygen.

5. *The Acids of Sulphur*

Sulphur forms a large number of different anions with oxygen, all of which are divalent. The sulphur ion itself is also divalent, but in aqueous solution the water changes it for the most part into the monovalent ion SH', although a certain number of divalent sulphur ions S'' must also be taken as being present, more especially when the aqueous solution is concentrated.

The solutions of sulphuretted hydrogen are very little dissociated, and then almost exclusively into $H^{\bullet}$ and SH'. This dissociation is lessened still further by the presence of other stronger acids, in proportion to the concentration of the hydrogen ions. The solvent action of acids upon certain metallic sulphides depends upon this (as given at p. 81), the action being greater the greater the concentration of the hydrogen ions. The solubility of the metallic sulphide in water also enters into the question here, as has likewise been already explained.

The odour of sulphuretted hydrogen renders its detection easy. Its presence can also be proved by the blackening which it produces upon a piece of filter paper moistened with acetate of lead solution. The quantitative estimation is made either by precipitating it as a metallic sulphide, or by measuring its reducing action; a solution of iodine is the most convenient to use in the latter case, the iodine being reduced to hydriodic acid. This volumetric estimation is very easy and accurate, and is therefore to be preferred to the other, care being taken to dilute largely and to guard against any possible escape of sulphuretted hydrogen during the operation.

The sulphur ions, as they exist in solutions of alkaline sulphides, give a beautiful violet coloration when a nitro-prusside is added, this being probably due to the production of a new anion. Even in the alkaline solution the colour is very evanescent, while it disappears at once if the solution is acid. The brown spot which a solution of alkaline sulphide produces on a silver coin is also characteristic; and the reaction is one of general application, seeing that all the oxygen salts of sulphur are reduced to sulphides when heated with a mixture of sodium carbonate and charcoal.

Of the oxygenated ions of sulphur that of sulphuric acid is the most important. It is both recognised and determined as the very insoluble barium salt. On account of the extremely slight solubility of the latter, it has a great tendency to come down as very fine powder, the adsorptive action of which may give rise to very considerable errors in quantitative estimations. The way to avoid this is to bring down the precipitate in as large grains as possible, *i.e.* to precipitate in a somewhat hot and acid solution. The solvent action of the acid can be compensated for by using an excess of precipitant. This carrying down of dissolved substance by the precipitate is most marked when ferric salts are present, the production of a "solid solution"[1] being assumed here. In such a case it is a wise precaution to first reduce the ferric salts to ferrous, which

[1] Cf. Van 't Hoff, *Zeitschr. für physik. Chemie.* **5**, 322 (1890).

are much less liable to be carried down. Or the sulphate of barium containing ferric oxide may be afterwards fused with alkaline carbonate, so as to produce barium carbonate and alkali sulphate, and the aqueous solution of the fused mass precipitated anew with barium salt. The iron remains in the insoluble residue.

In determining sulphuric acid we have sometimes to bear in mind the fact that it may be present as a complex compound, the compounds of chromium more particularly showing a tendency to this (cf. p. 145). Fusion with an excess of alkaline carbonate destroys the complex acids and converts them into sulphates.

Sulphurous acid is a far weaker acid than sulphuric, hence the sparingly soluble salts that it forms with barium, lead, etc., are soluble in acids. Its detection —apart from the odour—depends on the one hand upon the reducing actions which it shows, and on the other upon the proof of the sulphuric acid resulting from its oxidation. A third test is the reduction of sulphurous acid by nascent hydrogen, when sulphuretted hydrogen is formed, this last reaction being shared in common with all the other oxygen acids of sulphur excepting sulphuric. The ready formation of sulphuretted hydrogen appears to be a property of those compounds of sulphur which contain one atom of hydrogen linked to sulphur.

The reducing effect of sulphurous acid is strikingly shown in presence of iodic acid, when free iodine is liberated. Hydriodic acid is at first formed, and this at once acts upon some more of the iodic acid to produce iodine and water. Expressed in terms of the electric dissociation theory, iodine ions, together with those of iodic acid, cannot exist in presence of hydrogen ions;

they immediately pass into the non-ionised products iodine and water.

Sulphurous acid differs from the other oxygen acids of sulphur, about to be mentioned, by not giving any deposition of sulphur when treated with hydrochloric acid, but only liberating sulphur dioxide (as the others do also). Dithionic acid forms an exception, being decomposed under these circumstances into sulphur dioxide and sulphuric acid.

Free thiosulphuric acid is unknown, the salts only being capable of existence. It may be asked—Why should not the ion S_2O_3'' be as stable in an acid solution as in a neutral or alkaline one, seeing that we have to do with the same ion in all three cases? The answer is that this ion cannot exist alongside of hydrogen ions, since it is able to yield sulphur and sulphurous acid—*i.e.* non-ionised and more slightly ionised substances—with the latter. The reaction is not an ionic one, and therefore does not take place instantaneously; the time required for it depends upon the concentration of the hydrogen ions.

The thiosulphates find an important application in iodometric analysis. Two atoms of iodine pass here into two negative iodine ions, the necessary ion charges being taken from two S_2O_3'' ions, which lose two valencies and coalesce to the ion S_4O_6''.

The other halogens do not react in this way with the thiosulphates, but yield sulphuric acid and free sulphur. This difference in behaviour is to be traced to the fact that the tetrathionates too are oxidised by chlorine or bromine, giving sulphuric acid and sulphur. There appears to be no oxidation of the latter substance so long as an excess of thiosulphate is present, and hence the quantity of halogen may be calculated from the amount of sulphuric acid formed. It is, however, much simpler to allow the halogen to act upon iodide of potassium and then to titrate the separated iodine with sodium thiosulphate.

One of the two hydrogen atoms in thiosulphuric acid is linked directly to sulphur. It is therefore replaced with great readiness by the heavy metals, which have a strong affinity for sulphur, and the resulting compounds are but very slightly ionisable at this point (*i.e.* do not readily give up the metal as an ion). Hence many sparingly soluble metallic salts dissolve in an excess of thiosulphate by changing into complex anions which contain the metal linked to the sulphur, and in whose solutions there are extremely few metallic ions. From soluble metallic salts the thiosulphates generally first precipitate the sparingly soluble thiosulphate of the metal in question, and this then dissolves in the excess of thiosulphate to the salt of the metallo-thiosulphonic acid. The compounds of copper, lead, silver, mercury, etc., are examples in point. The thiosulphonates of the metals are by no means stable; most of them decompose in a neutral solution, and all of them in an acid one into metallic sulphide, sulphuric acid, sulphur, etc. This latter reaction is likewise applied in analysis to throw down the sulphides of the copper group without the use of sulphuretted hydrogen.

The phenomena which are dependent upon the presence of the atomic group SH are also shown by sulphurous acid, which likewise contains an atom of hydrogen linked to sulphur; but here they are less pronounced, and there is no transformation into metallic sulphide, because the acid contains only one atom of sulphur. Still, chloride of silver (for example) is almost as soluble in sodium sulphite as in the thiosulphate. Quite similar phenomena are to be expected from the polythionic acids, but there have apparently been no investigations made in this direction as yet. The formation of such complexes, in which the metallic ion is present for the most part in the non-ionised state, can be proved not merely by the solution of the sparingly soluble salts, but also by the measurement of the electromotive force of the metals in question in such solutions, for under those conditions the electrical position of the metal appears to be more or less displaced towards the zinc side.

6. *Carbonic Acid*

Carbonic acid is one of the weakest acids that still possess the true acid character. Its aqueous solution indeed shows an acid reaction, but it only changes blue litmus to a wine red, and not to a bright red as the stronger acids do. This arises partly from the small concentration that can be attained in an aqueous solution of carbonic acid, because of its sparing solubility at atmospheric pressure; when the solubility is increased by the application of a stronger pressure, the bright red coloration sets in.

Of the salts of carbonic acid only those of the alkalies are soluble in water; the metals of the alkaline earths form soluble bicarbonates, which are, however, very unstable, breaking up partially in the cold and completely when the liquid is boiled. The reason of this is that one of the decomposition products—the carbonic acid—is carried away by the vapour of the boiling water, so that the decomposition must go on until it is complete. The solutions of the alkaline carbonates have an alkaline reaction; the tendency which carbonic acid has to change into a less dissociated condition gives rise to the formation of acid carbonate, *i.e.* of HCO_3 ions, for which the requisite hydrogen must be abstracted from the water. The hydroxyl then remaining over causes the alkaline reaction.

The property which carbonic acid shows of not forming normal salts with weak bases is also connected with the marked weakness of the acid. Hydrolysis sets in, and the precipitate contains a mixture of carbonate and hydroxide, the proportion of the latter to the former increasing with the amount of water present. H. Rose made a series of extended observations upon this point a very long time ago (in 1857), and his results all fell out in the direction just indicated.

Carbonic acid is easily recognised by the readiness with which it passes into gaseous carbon dioxide, which escapes when almost any acid is added to a soluble or even insoluble carbonate. The strength of carbonic acid is so very slight that the influence of the "insolubility" in the latter instance is practically nothing; the decomposition of acetate of lead by carbonic acid is almost the only case of the kind that has been investigated to any extent. The qualitative test for carbonic acid is made with lime water, from which it precipitates carbonate of calcium. It is determined quantitatively either by absorption with soda-lime, or, when present only in very small amount, by receiving it in a measured volume of baryta-water of known strength, allowing the precipitate to settle, and then determining with standard acid the excess of baryta remaining in solution.

Carbonic acid is an invariable constituent of ordinary distilled water, into which it passes from the water originally taken. Part of it escapes when the water stands open to the air, but another part remains persistently behind. It may be displaced with tolerable completeness by passing a current of air free from carbon dioxide through the water for a long time. By this procedure the water remains purer than after boiling, as a very appreciable amount of material is usually dissolved from the glass in the latter case. A current of hydrogen is used if it is at the same time desired to prevent ingress of oxygen into the water.

7. *Phosphoric Acid*

In ortho-phosphoric acid, H_3PO_4, we find the influence exerted by the gradual ionisation of the hydrogen atoms of polybasic acids (which was discussed on p. 61), showing itself in a very marked degree. While the dissociation of the first hydrogen ion corresponds to that of an acid of medium strength, the second behaves like the ion of a weak acid, while the third is hardly capable of replacement at all in aqueous solution, the only soluble tri-metallic phosphates—those of the alkali metals and of ammonium—being broken up almost completely by hydrolysis into the bi-metallic phosphates (*i.e.* their ions) and free alkali. In other words, we find in the aqueous solution of the salt, Na_3PO_4, in addition to the sodium ions, not the trivalent ion PO_4''', but the divalent ion HPO_4'' and hydroxyl, OH'. The reason of this is that the tendency to ionisation of the third hydrogen atom is much smaller than that of the water; when, therefore, the salt Na_3PO_4 is dissolved in water, the ion PO_4''' immediately acts upon the latter, thus—

$$PO_4''' + \underbrace{H^{\cdot} + OH'} = PO_3H'' + OH'.$$

This difficulty of substitution is not shown in the case of solid and therefore of sparingly soluble salts, since these are not ionised. The hypothetical explanation of the phenomenon is that the development of a negative ion charge must be far more easy to bring about on a neutral atomic complex than on one which is already negatively charged, since the work required in the latter case must be much greater, other things being equal. And this applies still more to the development of the third ion charge. In solid non-ionised salts this condition is absent, and the normal tri-metallic phosphates are therefore perfectly stable in the solid state; they are, moreover, the only phosphates that occur in nature.

These relations are shown very clearly in the phenomenon attending the precipitation of ordinary sodium phosphate, Na_2HPO_4, by a silver solution, when the weakly alkaline reacting phosphate and the neutral silver salt yield a yellow precipitate of tri-argentic phosphate and a liquid of strongly acid reaction. The ordinary chemical equation

$$3AgNO_3 + Na_2HPO_4 = Ag_3PO_4 + 2NaNO_3 + HNO_3$$

gives only an imperfect idea of the reaction, for the latter is by no means complete, the liberated nitric acid dissolving some of the silver phosphate. The following equation is probably more correct, although it no doubt also fails to express all that goes on—

$$3AgNO_3 + 2Na_2HPO_4 = Ag_3PO_4 + 3NaNO_3 + H_2NaPO_4 \,;$$

according to this we do not get free nitric acid, but the acid-reacting dihydrogen-sodium phosphate (or rather its ions). The fact that the dihydrogen-sodium phosphate has an acid reaction is a sign that its solution contains hydrogen ions; these arise from the ionisation of the monovalent anion H_2PO_4' into hydrogen and the divalent anion HPO_4'', thus—

$$H_2PO_4' = HPO_4'' + H\cdot.$$

It is worthy of notice that the salts of orthophosphoric acid with trivalent cathions like aluminium, iron and chromium are very slightly soluble indeed. There is apparently a general law underlying this phenomenon, according to which compounds that are built up of ions of equal valency have a special tendency to form sparingly soluble salts. The typical precipitants for the pronounced monovalent halogens are the monovalent cathions of silver, mercury and copper; for the divalent alkaline earth metals the divalent ions of sulphuric, oxalic and carbonic acids serve as precipitating agents; while in the case of the trivalent ions of iron, chromium and aluminium, the phosphates are insoluble in acetic acid, which dissolves the other sparingly soluble salts of these metals. The law cannot, however, be reversed; for, although the most insoluble compounds throughout are made up of equi-valent ions, there are on the other hand numerous salts with ions of the same valency which dissolve readily in water. There is thus obviously some other condition exerting an influence on solubility which obscures in many cases the regularity just referred to, but what this may be I do not know.

Phosphoric acid is capable of forming complex compounds with various metallic acids, more especially with tungstic and molybdic, in which the basicity of the acid remains constant while the molecular proportions of the trioxides in question vary.

Phospho-molybdic acid, which is the most important example of this type from an analytical point of view, forms very sparingly soluble yellow salts with the alkali metals and with ammonium, which dissolve to but a very slight extent even in free acid, especially when one of their ions is present in excess. This is made use of for the detection and quantitative separation of phosphoric acid from nitric acid solutions, an excess of solution of molybdic acid and ammonium nitrate in nitric acid being added to the liquid under examination. The mixed solution has to be warmed gently and allowed to stand for a considerable time before the reaction completes itself. Here, again, we have a reaction which is not purely ionic, and hence it requires a measurable time. The reaction could be followed quantitatively by investigating the electric conductivity, the specific volume, the colour, or any other convenient property of the solution.

The complex ions of phospho-molybdic acid are stable only in acid solution, being broken up by excess of alkali or ammonia into salts of phosphoric and molybdic acids. The yellow precipitate thus dissolves readily in ammonia, and the phosphoric acid can be completely thrown down from the solution as ammonium-magnesium phosphate. This process is extensively followed, not merely in analysis, but also for freeing the molybdenum residues from phosphoric acid, with the object of using the molybdic acid over again.

Ortho-phosphoric acid changes into pyro-phosphoric acid, $H_4P_2O_7$, and meta-phosphoric acid, HPO_3, by giving up the elements of water. The meta-acid is not a true analogue of nitric acid, as one might have expected from the connection between nitrogen and phosphorus, but is like pyro-phosphoric acid a condensed acid of materially higher molecular weight than is indicated by the formula HPO_3. There are indeed a number of different meta-phosphoric acids of different molecular weights and varying properties. The meta-acid present in fused or vitreous phosphoric acid possesses the property of precipitating albumen from solution, and it also gives a white silver salt. Pyro-phosphoric acid does not precipitate albumen, but yields a precipitate with chloride of barium, which the ortho-acid does not do. Neither the meta- nor the pyro-acid shows the reactions of the ortho-acid with magnesia mixture or molybdate of ammonium.

It is important from an analytical point of view that these derivatives of phosphoric acid pass into ortho-phosphate or ortho-phosphoric acid when they are fused with an excess of alkaline carbonate or warmed for a length of time in a strongly acid solution. The change takes place with the acids themselves on merely letting their aqueous solutions stand; the dry salts can, however, be preserved unaltered. For a quantitative estimation the pyro- and meta-phosphoric acids are always converted into the ortho-compound, which is then precipitated as magnesium-ammonium phosphate (cf. p. 137).

8. *Phosphorous and Hypophosphorous Acids*

Although these two acids are di- and monobasic respectively, they may be discussed here in connection with phosphoric acid, since they are always transformed into the latter for quantitative estimation.

The simplest analytical characteristic of the lower acids of phosphorus is the liberation of spontaneously inflammable phosphuretted hydrogen when either they themselves or their salts are heated; at the same time red phosphorus is separated. They also act as reducing agents, precipitating (*e.g.*) calomel from an acid solution of corrosive sublimate. For the rest they form mostly soluble salts (barium phosphite is sparingly soluble in water though readily in acids) in which there is little that is uncommon.

When these acids come into contact with nascent hydrogen they are reduced to phosphuretted hydrogen, while phosphoric acid is not. This is exactly the same behaviour that we have already found in the case of sulphuric acid and the lower acids of sulphur, and it stands in close connection with the way in which the hydrogen is linked in the acids. For in all probability the formula $OP{\displaystyle{-OH\atop-OH}\atop OH}$ expresses the constitution of phosphoric acid, while the other acids have the formulæ $OP{\displaystyle{-OH\atop-OH}\atop H}$ and $OP{\displaystyle{-OH\atop-H}\atop H}$. There is thus no hydrogen linked directly to phosphorus in the first of these, but there is in the others.

When the acids are pure they can be distinguished from one another by the fact that hypophosphorous acid shows with an indicator the sharp point of neutralisation upon the gradual addition of alkali that is characteristic of a fairly strong monobasic acid. The dibasic phosphorous acid on the other hand is marked by the same peculiarity as the tribasic phosphoric, in that the second hydrogen atom is much more difficult to substitute in aqueous solution than the first, so that their neutral salts are broken up hydrolytically to a certain extent and give an alkaline reaction. If, therefore, an acid liquid which contains the lower acids of phosphorus shows a sharp point of neutralisation, only hypophosphorous acid is present; but if the change in colour is not sharp, then the liquid contains phosphorous acid (other acids which show indistinct neutralisation, especially phosphoric, being of course absent), although hypophosphorous acid may also be present at the same time.

All the reducing actions effected by the lower acids of phosphorus go on at an exceptionally slow rate.

9. *Boracic Acid*

Salts of normal boracic acid, H_3BO_3, are hardly known, for it shares with other weak acids the tendency to form condensed acids by the elimination of the elements of water from several molecules of the acid, the residues coalescing to a more complex compound. The best known of these polybasic acids is the tetraboracic, $H_2B_4O_7$, the acid of ordinary borax. There are no differences apparent between the various borates in aqueous solution, so far as the reactions of the boracic acid itself are concerned; sharp and clear reactions are wanting here.

Boracic acid is easily recognised by the green colour which it imparts to the flame of burning alcohol. There is a marked distinction between this coloured flame and those of the alkalies, for instance, as it is not necessary that the flame-colouring substance should be raised to a white heat in this case. The boracic acid volatilises with the vapour of the boiling alcohol by forming a volatile boracic ether. The test is best made by covering the substance in a small crucible with concentrated sulphuric acid, adding plenty of alcohol, and then warming until the latter boils and takes fire; if the flame should be coloured green under those circumstances, this can only be due to boracic acid.

Turmeric paper furnishes another very delicate test for boracic acid, the original yellow colour being changed to a reddish-brown after the paper has been dipped into a solution of the acid and dried at a gentle heat. Nothing is known as to the cause of this curious reaction, but it may possibly be dependent upon the following property :—

Boracic acid possesses the peculiarity of forming complex acids with organic compounds containing several hydroxyls in their molecule, which show a far more acid reaction than either boracic acid itself or the organic compounds in question. Probably the monovalent radicle boracyl, BO, takes the place of the hydroxyl hydrogen here, as in the corresponding antimony compounds.

10. *Silicic Acid*

Silicic is an extraordinarily weak acid, the only soluble salts that it forms being those with the alkali metals. The aqueous solutions of these are split up hydrolytically to a very great extent, so that they show a strongly alkaline reaction; the free silicic acid contained in such solutions is not dissolved in the ordinary form, but is in the colloidal state, and has consequently very little power of reaction. The changes in equilibrium brought about by a dilution or concentration of this solution do not therefore follow instantaneously, but require a more or less considerable time, and the phenomenon of *chemical after-effect* is very pronounced here, in that solutions of the same composition and at the same temperature possess by no means the same properties, but different ones, according to the conditions through which the solution has passed. The electrical conductivity furnishes the best means of observing such differences.

The silicates of the other metals are insoluble in water; some of them can be decomposed by the ordinary mineral acids, but others are not affected. Speaking generally, the more basic a silicate is the easier it is to decompose, and hence it is a rule in analysis to bring about this decomposability through acids by first fusing the silicate with an excess of the carbonates of potassium and sodium. By this means all the constituents of the silicate excepting the alkalies can be determined; for the estimation of the latter, the silicate must be treated with hydrofluoric acid. To this end the finely powdered mineral is covered with an excess of aqueous hydrofluoric acid, and the whole evaporated to dryness after the addition of sulphuric acid. The silicon escapes as fluoride, while the metals remain behind as sulphates. The addition of sulphuric acid is essential, because silicon fluoride is decomposed by water, and hence some hygroscopic substance must be present in order that the volatilisation may be complete.

When silicates are decomposed by acids, the silicic acid is separated in the colloidal state. According, therefore, to the degree of concentration, it either remains apparently dissolved (when the solution is very dilute), or it separates out in the form of jelly or powder. It is then at least partly soluble, whatever the conditions may be; so, to render it completely insoluble, the whole must be evaporated to dryness, and the residue heated for some time to a little above 100°. It is necessary to take up this residue with dilute hydrochloric acid and not with water alone, as otherwise we get basic chlorides of magnesium, iron and aluminium formed, which are not perfectly soluble in water.

The qualitative test for silicic acid depends upon its insolubility in a bead of fused sodium metaphosphate. The metals which are combined with the silica to silicates dissolve in this, leaving — if silicic acid is present—a siliceous "skeleton," *i.e.* the undissolved silica swims about in the fused bead.

Silicic acid hardly shows any ion-reactions proper, and at any rate none are applied for purposes of analysis.

CHAPTER XIII

THE CALCULATION OF ANALYSES

SINCE as a general rule the substances separated by analysis or otherwise quantitatively determined are not identical with those whose percentages it is the aim of the analyst to elucidate (cf. p. 100), the results obtained in the first instance have to undergo subsequent calculation. According to the fundamental stœchiometric laws, the amounts of substances which are capable of transformation one into the other are proportional among themselves, and therefore the calculation just referred to merely consists in multiplying by a definite factor, which represents the ratio between the combining weight of the substance found and of that required. In this way we arrive at the amount of the particular constituent in the material under analysis. The results are usually reckoned upon 100 parts of the original substance, so that the final numbers represent percentages of the various ingredients.

With respect to the calculation of the ultimate constituents, there is a total want of agreement between the various branches of chemistry. The most rational method is that followed in organic chemistry, where the calculation is always made back to the constituent elements themselves, and the results of the analysis stated in this way without reference to any views that may be held regarding the constitution of the compound analysed. In inorganic chemistry, on the other hand, there is the greatest discrepancy in this respect. While the results of the analysis of compounds of entirely unknown constitution or of mixtures are often given in percentages of the constituent elements, it is usual in the case of compounds whose constitution is known, or is supposed to be known, to group the elements into proximate constituents. This naturally allows free play for the advancement of special views and practical considerations of the most various kinds, and there are in fact some procedures still in vogue here which have been altogether abandoned in the other branches of the science.

The department of mineral analysis affords us a striking example in point. It is still customary, in stating the composition of a complex silicate, to adhere to the dualistic formulæ of Berzelius, and to give the metals as oxides and the acids as anhydrides. The reason for this ultra-conservative procedure obviously lies in the fact that the arithmetical control of the results is most easily attained by so doing, since the sum of the constituents thus calculated must be equal to the original amount of substance taken, or—in a percentage calculation—equal to 100. This advantage however disappears when halogens are present in the compound, since their hydracids—which contain no oxygen—cannot be formulated as anhydrides. The analyst often helps himself out in such a case by imagining the halogen combined with one of the metals present and calculating it accordingly, although such a procedure is necessarily an arbitrary one.

The calculation is still more arbitrary after an analysis of a mixture of dissolved salts, such as occur in natural waters. With regard to this, efforts had long been made in vain to find some definite basis for the answer to the question—How are the bases and acids present combined with one another? The final answer, to which we are led by the dissociation theory, is that these are not combined at all, but that they—or rather the ions of the salts—lead separate existences, to which the only limitation is the law that the sum total of the positive ions must be equivalent to the sum total of the negative.

It follows from this that the simplest and best way of stating the results of an analysis would be to give only the ultimate elements themselves and their relative amounts, and I do not hesitate to recommend this procedure as being the most correct in principle. Of course, by doing this one could not show, in the analytical statement of results, in what form the various elements were present in the compound; but it seems to me that it would be more appropriate to give such particulars separately, and thus to keep the actual results of the analysis altogether free from any hypotheses. It is true that we can in many cases cite experiment in justification of the old procedure, for instance, when a compound contains iron both in the ferrous and ferric states; but it is easy to indicate this by some suitable sign, by Fe·· and Fe··· in the example just mentioned.

Another case in which one would prefer to bring groups into the calculation instead of merely the elements alone, would be that referred to in the last paragraph but one, when we know that the substance analysed is a mixture of neutral salts such as occur in sea water and other similar natural solutions. We learn (*e.g.*) from this analysis that sulphur is not only present in the solution, but that it is present as sulphate in the form of the ion SO_4. In such a case it is best to give the relative amounts of the ions without attempting to combine them together, as is still done even now, notwithstanding the definite statement in text-books that we know nothing for certain about the determining causes upon which this combination depends. A certain difficulty is caused here by carbonic acid, if it is present in excess, as it generally is in spring and well waters. The simplest plan is to calculate the "combined" carbonic acid as CO_3 (which is the ion of the normal carbonates) from the amount of the metallic ions after deducting the other anions; whatever carbonic acid remains over is to be given as free carbonic anhydride, CO_2. This is not in truth quite correct, for it is tolerably certain that such solutions in which excess of carbonic acid is present do not contain the ion CO_3, but the monovalent ion HCO_3 of the acid carbonates. Since, however, these are converted more or less completely into normal carbonates by boiling, it seems permissible to disregard this small complication and to reckon the carbonates as normal.

The same rules would apply to all other such cases in which one was justified in laying stress upon a knowledge of the ions present.

THE END

Printed by R. & R. Clark, Limited, *Edinburgh*

MESSRS. MACMILLAN AND CO.'S BOOKS

FOR

STUDENTS OF CHEMISTRY.

A TREATISE ON CHEMISTRY. By Sir H. E. ROSCOE, F.R.S., and the late C. SCHORLEMMER, F.R.S. Vol. I. The Non-Metallic Elements. New Edition. Completely revised by Sir H. E. ROSCOE, assisted by Drs. H. G. COLMAN and A. HARDEN. With 374 Illustrations and a Portrait of Dalton engraved by C. H. JEENS. 8vo. 21s. Vol. II. Part I. Metals. 18s. Vol. II. Part II. Metals. 18s. Vol. III. Organic Chemistry. Parts I., II., IV., and VI. 21s. each. Parts III. and V. 18s. each.

INORGANIC CHEMISTRY FOR BEGINNERS. By Sir HENRY E. ROSCOE, F.R.S. Assisted by JOSEPH LUNT, B.Sc. Globe 8vo. 2s. 6d.

LESSONS IN ELEMENTARY CHEMISTRY, INORGANIC AND ORGANIC. By Sir H. E. ROSCOE, F.R.S. Sixth Edition, thoroughly Revised. Fcap. 8vo. 4s. 6d.

ANALYTICAL CHEMISTRY. By A. MENSCHUTKIN, Professor in the University of St. Petersburg. Translated from the third German Edition, under the supervision of the Author, by JAMES LOCKE. Medium 8vo. 17s. net.

THEORETICAL CHEMISTRY FROM THE STANDPOINT OF AVOGADRO'S RULE AND THERMO-DYNAMICS. By Prof. NERNST. Translated by Prof. CHARLES SKEELE PALMER, of the University of Colorado. With 26 Woodcuts and Two Appendices. 8vo. 15s. net.

THE RISE AND DEVELOPMENT OF ORGANIC CHEMISTRY. By CARL SCHORLEMMER, LL.D., F.R.S., late Professor of Organic Chemistry in the Owens College, Manchester, Victoria University. Revised Edition. Edited by ARTHUR SMITHELLS, B.Sc. Crown 8vo. 5s. net.

A LABORATORY MANUAL OF ORGANIC CHEMISTRY. By Dr. LASSAR COHN, Professor of Chemistry in the University of Konigsberg. Translated, with the Author's sanction, from the second German Edition, by ALEXANDER SMITH, B.Sc., Ph.D. Crown 8vo. 8s. 6d.

ESSAYS IN HISTORICAL CHEMISTRY. By T. E. THORPE, F.R.S., Professor of Chemistry in the Royal College of Science, South Kensington, London. Crown 8vo. 8s. 6d. net.

HISTORY OF CHEMISTRY FROM THE EARLIEST TIMES TO THE PRESENT DAY. By ERNST VON MEYER, Ph.D. Translated by GEORGE M'GOWAN, Ph.D. 8vo. 14s. net.

EXPERIMENTAL PROOFS OF CHEMICAL THEORY FOR BEGINNERS. By WILLIAM RAMSAY, Ph.D., F.R.S., Professor of Chemistry in University College, London. Second Edition. Pot 8vo. 2s. 6d.

CHEMICAL THEORY FOR BEGINNERS. By LEONARD DOBBIN, Ph.D., and JAMES WALKER, Ph.D., D.Sc. Pot 8vo. 2s. 6d.

OUTLINES OF GENERAL CHEMISTRY. By Prof. WILHELM OSTWALD. Translated by J. WALKER, Ph.D., D.Sc. 8vo. 10s. net.

MANUAL OF PHYSICO-CHEMICAL MEASUREMENTS. By Prof. WILHELM OSTWALD. Translated by JAMES WALKER, D.Sc., Ph.D. Illustrated. 8vo. 7s. net.

A TEXT-BOOK OF INORGANIC CHEMISTRY. By Prof. IRA REMSEN. 8vo. 16s.

INORGANIC CHEMISTRY. By the same. Crown 8vo. 6s. 6d.

ORGANIC CHEMISTRY. By the same. Crown 8vo. 6s. 6d.

THE ELEMENTS OF CHEMISTRY. By the same. New Edition. Fcap. 8vo. 2s. 6d.

MACMILLAN AND CO., LONDON.

MESSRS. MACMILLAN AND CO.'S BOOKS
FOR
STUDENTS OF CHEMISTRY.

PRACTICAL INORGANIC CHEMISTRY. By FRANCIS JONES, F.R.S.E. Fcap. 8vo. 2s. 6d.

PRACTICAL ORGANIC CHEMISTRY. By J. B. COHEN, Ph.D., Fcap. 8vo. 2s. 6d.

LESSONS IN ORGANIC CHEMISTRY. Part I., Elementary. By G. S. TURPIN, M.A., D.Sc., Principal of the Technical School, Huddersfield. 2s. 6d.

PRACTICAL INORGANIC CHEMISTRY. By G. S. TURPIN, M.A., D.Sc. Adapted to the New Syllabus (1895) of the Science and Art Department. Globe 8vo. 2s. 6d.

A MANUAL OF MICROCHEMICAL ANALYSIS. By Prof. H. BEHRENS. With an Introductory Chapter by Prof. J. W. Judd, F.R.S. 6s.

CHEMICAL ANALYSIS OF OILS, FATS, AND WAXES, AND OF THE COMMERCIAL PRODUCTS DERIVED THEREFROM. From the German of Prof. Dr. R. BENEDIKT. Revised and enlarged by Dr. J. LEWKOWITSCH, F.I.C., F.C.S., formerly Technical Manager at the Whitehall Soap Works, Leeds; Consulting Chemist and Chemical Engineer. 8vo. 21s. net.

A SERIES OF CHEMICAL PROBLEMS WITH KEY. By T. E. THORPE, F.R.S. New Edition. Revised by W. TATE. Fcap. 8vo. 2s.

CHEMICAL ARITHMETIC. With 1200 Problems. By S. LUPTON, M.A. Second Edition, revised. Fcap. 8vo. 4s. 6d.

THE ELEMENTS OF THERMAL CHEMISTRY. By M. M. PATTISON MUIR, M.A., F.R.S.E., assisted by D. M. WILSON. With a Plate. 8vo. 12s. 6d.

A TEXT-BOOK OF THE PHYSIOLOGICAL CHEMISTRY OF THE ANIMAL BODY, including an Account of the Chemical Changes occurring in Disease. By ARTHUR GAMGEE, M.D., F.R.S. Vol. II. The Physiological Chemistry of Digestion. With two Chromo-Lithographic Charts by SPILLON and WILKINSON. 8vo. 18s. (Vol. I. 8vo. 18s.)

THE CHEMICAL BASIS OF THE ANIMAL BODY. An Appendix to Foster's "Text-Book of Physiology." Fifth Edition. By A. SHERIDAN LEA, M.A., D.Sc., F.R.S., University Lecturer in Physiology in the University of Cambridge; Fellow of Caius College. 8vo. 7s. 6d.

PRIMER OF AGRICULTURAL CHEMISTRY; OR, THE FOOD OF PLANTS. By A. P. LAURIE, M.A., Fellow of King's College, Cambridge, Adviser in Technical Education to the Bedfordshire County Council. Pot 8vo. 1s.

BLOWPIPE ANALYSIS. By J. LANDAUER. Authorised Translation by J. TAYLOR, of Owens College, Manchester. New and thoroughly revised Edition. Globe 8vo. 4s. 6d.

METHODS OF GAS ANALYSIS. By Dr. WALTHER HEMPEL. Translated by Dr. L. M. DENNIS. Crown 8vo. 7s. 6d.

THE CHEMISTRY OF PHOTOGRAPHY. By RAPHAEL MELDOLA, F.R.S., Professor of Chemistry, Technical College, Finsbury. Crown 8vo. 6s.
[*Nature Series.*

MACMILLAN AND CO., LONDON.

MESSRS. MACMILLAN AND CO.'S BOOKS
FOR
STUDENTS OF PHYSICS.

A TREATISE ON BESSEL FUNCTIONS AND THEIR APPLICATIONS TO PHYSICS. By A. GRAY, M.A., F.R.S.E., and G. B. MATHEWS, M.A., Fellow of St. John's College, Cambridge. 8vo. 14s. net.

SIR ISAAC NEWTON'S PRINCIPIA. Reprinted for Lord KELVIN, P.R.S., and HUGH BLACKBURN, M.A. 4to. 31s. 6d.

NEWTON'S PRINCIPIA. Sections I., II., III. With Notes and Illustrations. Also a Collection of Problems principally intended as Examples of Newton's Methods. By PERCIVAL FROST, D.Sc., F.R.S., Fellow and Mathematical Lecturer, King's College, Cambridge. 8vo. 12s.

AN ESSAY ON THE GENESIS, CONTENTS, AND HISTORY OF NEWTON'S "PRINCIPIA." By WALTER W. ROUSE BALL, Fellow and Assistant Tutor of Trinity College, Cambridge. Crown 8vo. 6s. net.

UTILITY OF QUATERNIONS IN PHYSICS. By ALEXANDER M'AULAY, M.A., Lecturer in Mathematics and Physics in the University of Tasmania. 8vo. 5s. net.

AN ELEMENTARY TREATISE ON THEORETICAL MECHANICS. By Professor ALEXANDER ZIWET, Assistant Professor of Mathematics in the University of Michigan. Part I. Kinematics. Part II. Introduction to Dynamics; Statics. 8vo. 8s. 6d. net each part.

APPLICATIONS OF DYNAMICS TO PHYSICS AND CHEMISTRY. By J. J. THOMSON, M.A., F.R.S., Fellow of Trinity College, and Cavendish Professor of Experimental Physics, Cambridge. Crown 8vo. 7s. 6d.

A TREATISE ON THE MOTION OF VORTEX RINGS. An Essay to which the Adams Prize was adjudged in 1882, in the University of Cambridge. By J. J. THOMSON, F.R.S. 8vo. 6s.

THE ELEMENTS OF GRAPHIC STATICS. A Text-Book for Students of Engineering. By L. M. HOSKINS, Professor of Pure and Applied Mathematics in the Leland Stanford Junior University, formerly Professor of Mechanics in the University of Wisconsin. 8vo. 10s. net.

A TREATISE ON HYDROSTATICS. By ALFRED GEORGE GREENHILL, F.R.S., Professor of Mathematics in the Artillery College, Woolwich. Crown 8vo. 7s. 6d.

A TEXT-BOOK OF THE PRINCIPLES OF PHYSICS. By ALFRED DANIELL, M.A., Lecturer on Physics in the School of Medicine, Edinburgh. Third Edition. Medium 8vo. 21s.

PHYSICS. Advanced Course. By GEORGE F. BARKER, Professor of Physics in the University of Pennsylvania. With a Magnetic Map of the United States. 8vo. 21s.

MACMILLAN AND CO., LONDON.

MESSRS. MACMILLAN AND CO.'S BOOKS

FOR

STUDENTS OF PHYSICS.

LESSONS IN ELEMENTARY PRACTICAL PHYSICS. By BALFOUR STEWART, LL.D., F.R.S., and W. W. HALDANE GEE, Demonstrator and Assistant Lecturer in Physics, the Owens College. Vol. I. General Physical Processes. Crown 8vo. 6s. Vol. II. Electricity and Magnetism. Crown 8vo. 7s. 6d.

AN EXERCISE BOOK OF ELEMENTARY PRACTICAL PHYSICS, for Organised Schools under the Department of Science and Art Evening Continuation Schools, and Elementary Day Schools, arranged according to the Headmasters' Association Syllabus of Practical Physics. By RICHARD A. GREGORY, F.R.A.S. Fcap. 4to.

LECTURES ON SOME RECENT ADVANCES IN PHYSICAL SCIENCE. By P. G. Tait, M.A., Sec. R.S.E., formerly Fellow of St. Peter's College, Cambridge; Professor of Natural Philosophy in the University of Edinburgh. Third Edition. Crown 8vo. 9s.

HEAT. By Prof. P. G. TAIT. Crown 8vo. 6s.

THE THEORY OF HEAT. By THOMAS PRESTON, M.A. (Dub.), Fellow of the Royal University of Ireland, and Professor of Natural Philosophy, University College, Dublin. 8vo. 17s. net.

BURNETT LECTURES—ON LIGHT. In Three Courses. Delivered at Aberdeen in November 1883, December 1884, and November 1885. By Sir GEORGE GABRIEL STOKES, M.A., F.R.S., Fellow of Pembroke College, and Lucasian Professor of Mathematics in the University of Cambridge. Second Edition. Crown 8vo. 7s. 6d. [*Nature Series.*

ELECTRIC WAVES. Being Researches on the Propagation of Electric Action with Finite Velocity through Space. By HEINRICH HERTZ, late Professor of Physics in the University of Bonn. Authorised translation by D. JONES, B.Sc. With Preface by Lord KELVIN, P.R.S. Illustrated. 8vo. 10s. net.

PAPERS ON ELECTROSTATICS AND MAGNETISM. By Lord KELVIN, P.R.S. Second Edition. 8vo. 18s.

MODERN VIEWS OF ELECTRICITY. By OLIVER J. LODGE, D.Sc., LL.D., F.R.S., Professor of Experimental Physics in University College, Liverpool. With Illustrations. Crown 8vo. 6s. 6d. [*Nature Series.*

THE THEORY AND PRACTICE OF ABSOLUTE MEASUREMENTS IN ELECTRICITY AND MAGNETISM. By ANDREW GRAY, M.A., F.R.S.E., Professor of Physics in the University College of North Wales. In two vols. Crown 8vo. Vol. I., 12s. 6d.; Vol. II., in two parts. 25s.

ABSOLUTE MEASUREMENTS IN ELECTRICITY AND MAGNETISM. By Professor ANDREW GRAY. Second Edition. Fcap. 8vo. 5s. 6d.

A TEXT-BOOK ON ELECTRO-MAGNETISM AND THE CONSTRUCTION OF DYNAMOS. By DUGALD C. JACKSON, B.S., C.E., Professor of Electrical Engineering, University of Wisconsin. 8vo. 9s. net.

MACMILLAN AND CO., LONDON.

Messrs. Macmillan & Co.'s
CLASSIFIED CATALOGUE
OF
Text-books and Works of Reference
FOR
Medical Students
AND
Practitioners

INDEX TO AUTHORS.

NAME.	PAGE.	NAME.	PAGE.
Alcock, T. (Huxley)	36	Earl, A.	15, 25
Allbutt, T. C.	41, 42	Edmunds, W. (Ballance)	41
Balfour, F. M.	27, 30, 31	Everett, J. D.	22
" " (Foster)	30	Fayrer, Sir J.	45
Ballance, C. A.	41	Fessenden, C. E.	15
Barker, G. F.	22	Fleischer, E.	9
Barnard, C. (Mayer)	18	Fletcher, H. M. (Kahlden)	37
Barnes, C. L. (Stewart)	20	Flower, Sir W. H.	30
Bateson, W.	27	Foster, M.	30, 34, 35
Bedell, F. (Nichols)	24	Franklin, W. S. (Nichols)	19
Bernard, H. and M. (Lang)	32	Gadow, H. (Flower)	30
" " (Wiedersheim)	32	Gamgee, A.	36
Berners, J.	43	Gattermann, L.	9
Bettany, G. T.	25	Gee, W. W. H. (Stewart)	20, 21
Biedermann, W.	34	" " (Wrapson)	22
Blyth, A. W.	44	Goodale, G. L	30
Bower, F. O.	25, 28	Gray, Andrew	23
Brunton, T. L.	39, 40	Gray, Asa	29
Cohen, J. B.	3	Gregory, R. A.	16
Comey, A. M.	8	Griffiths, W. H.	40
Copeman (S. M.)	46	Haeckel, E. (Lang)	32
Core, T. H. (Stewart)	19	Hamilton, D. J.	37
Corfield, W. H.	44	Harmer, S. F.	28
Crosskey, R. (Miers)	45	Hartley, W. N.	3
Daniell, A.	15, 22	Haswell, W. A. (T. J. Parker)	30
Dennis, L. M. (Hempel)	9	Hawkins, H. P.	43
Dobbin, L.	3	Heape, W. (Foster)	30
Drysdale, J. H. (Kanthack)	37	Hempel, W.	9

NAME.	PAGE.	NAME.	PAGE.
Henslow (Oliver)	26	Ostwald, W.	11, 12
Hooker, Sir J. D.	25, 29	Palmer, C. S. (Nernst)	11
Hotchkiss, H. G. (Nichols)	24	Parker, T. J.	27, 30, 32
Howes, G. B.	26	Parker, W. N. (Wiedersheim)	33
,, ,, (Huxley)	26	Parkes, L. C. (Corfield)	44
,, ,, (Wiedersheim)	32, 33	Perkin, W. H.	5
Huxley, T. H.	26, 36	Playfair, W. S. (Allbutt)	41
,, ,, (Howes)	26	Ramsay, W.	5
Jones, C.	5, 7	Remsen, I.	6, 7, 12
Jones, D. E.	16, 17	Reynolds, E. S.	45
Jones, F.	4	Rintoul, D.	16
Kahlden, C. v.	37	Rogers, F. J. (Nichols)	18
Kanthack, A. A.	37	Roscoe, Sir H. E.	7, 13, 14
Kimber, D. C.	36	,, ,, (Jones)	4
Klein, E.	38	,, ,, (Thorpe-Tate)	14
Lang, A.	32	Schorlemmer, C.	14
Langley, J. N. (Foster)	35	,, ,, (Roscoe)	13
Lassar-Cohn	9	Schuster, A.	24
Lea, A. S. (Foster)	35	,, ,, (Roscoe)	13
Lean, B. (Perkin)	5	Scott, D. H. (Huxley)	26
Le Blanc, M.	10	Sedgwick, A. (Foster)	30
Lee, F. S. (Verworn)	37	Sheild, M.	41
Lees, C. H. (Schuster)	24	Shipley, A. E. (Harmer)	28
Letts, E. A.	4	Shober, W. B. (Gattermann)	9
Locke, J. (Menschutkin)	10	Shore, L. E. (Foster)	35
Lodge, O. J.	23	Smith, A. (Lassar-Cohn)	9
Loewy, B.	17	Smithells, A. (Schorlemmer)	14
Loudon, W. G.	23	Stewart, B.	19, 21
Lunt, J. (Roscoe)	7	Stone, W. H.	21
Lupton, S.	17	Tate, W. (Thorpe)	14
MacAlister, D. (Ziegler)	38	Thompson, S. P.	21
Macdonald, G.	41	Thorne, R. Thorne	46
Martin, H. N. (Huxley)	26	Thorpe, T. E.	14
Matthews, C. P. (Nichols)	24	Tubby, A. H.	42
Mayer, A. M.	18	Turpin, G. S.	8
McLennan, G. C. (Loudon)	23	Verworn, M.	37
Menschutkin, A.	10	Walker, J. (Dobbin)	3
Merritt, E. (Nichols)	18	,, (Ostwald)	11
Meyer, E. v.	10	Watts, H. (Wurtz)	14
M'Gowan, G. (Ostwald)	12	Welby, F. A. (Biedermann)	34
,, ,, (Meyer)	11	White, W. H.	40
Miers, H. A.	45	Whitney, W. R. (Le Blanc)	10
Minot, C. S.	32	Wiedersheim, R.	32
Mivart, St. G.	31	Williams, F. H. (Brunton)	39
Mixter, W. G.	5	Willoughby, E. F.	46
Moler, G. S. (Nichols)	24	Wilson, D. M. (Muir)	11
Muir, M. M. Pattison	11	Woodhead, G. S. (Kahlden)	37
,, ,, (Fleischer)	9	Wrapson, J. P.	22
Nernst, W.	11	Wurtz, Ad.	14
Nichols, E. L.	18, 19, 24	Ziegler, E.	38
Oliver, D.	26		

CHEMISTRY.

ELEMENTARY.

Fcap. 8vo. 2s. 6d.

The Owens College Course of Practical Organic
Chemistry. By JULIUS B. COHEN, Ph.D., F.C.S., Assistant Lecturer on Chemistry, Owens College, Manchester.

THE ACADEMY.—"Dr. Cohen describes the preparation of eighty-seven organic bodies in this admirably planned series of laboratory lessons. The directions are clear and concise, while the illustrative woodcuts adequately represent the form and disposition of the apparatus required in the several operations. . . . Besides foot-notes, there are several pages of explanatory observations on the chief reactions described in the body of the work. This course of lessons will be found very useful in laboratories where advanced organic chemistry is taught. We know of no volume of similar scope."

Fcap. 8vo. 2s. 6d.

Chemical Theory for Beginners. By LEONARD
DOBBIN, Ph.D., Assistant in the Chemistry Department, University of Edinburgh, and JAMES WALKER, Ph.D., D.Sc., Professor of Chemistry in University College, Dundee.

THE JOURNAL OF EDUCATION.—"Elementary students are told just sufficient to enable them to work out a chemical problem. . . . This book is intended to come to the help of the student in his transition stage, and contains an exceptionally clear and concise explanation of chemical theory."

THE EDUCATIONAL REVIEW.—"We know of no other book which will serve so well to supplement the ordinary treatises. . . . The manner in which the joint Authors have carried out their plan is very satisfactory."

Globe 8vo. 5s.

A Course of Quantitative Analysis for Students.
By W. N. HARTLEY, F.R.S., Professor of Chemistry and of Applied Chemistry, Science and Art Department, Royal College of Science, Dublin.

THE BRITISH MEDICAL JOURNAL.—"This is a carefully executed little book, which will form in many respects a valuable laboratory outline to students engaged in quantitative chemical analysis. . . . The examples are well selected. . . ."

THE ATHENÆUM.—"The descriptions are very clear and great attention seems to have been paid to those little points of manipulation over which students generally come to grief; the notes following many of the processes and explaining the reasons for particular precautions are a very commendable feature in the book."

Fcap. 8vo.   2s. 6d.

The Junior Course of Practical Chemistry.
By FRANCIS JONES, F.R.S.E., F.C.S., Chemical Master in the Grammar School, Manchester, with a Preface by Sir HENRY E. ROSCOE, F.R.S. With Illustrations. New Edition.

THE JOURNAL OF EDUCATION.—"Though the general plan of this work is the same as in the previous editions, much fresh matter has been added. In the preparation of gases, etc., which forms the first part of the book, a greater number of reactions are given, and the preparation of marsh gas and ethylene is included. The analytical tables have also been expanded, a noteworthy addition being a table for determining bases in the presence of phosphates, borates, oxalates, etc., and the detection of organic acids in mixtures. . . . A special feature of this edition, however, is Part VII, which is entirely new, and consists of examples in simple qualitative analysis, to meet the requirements of the syllabus of the Science and Art Department, and also the preparation of some simple salts as required by the conjoint Board of Physicians and Surgeons, London."

THE GUARDIAN.—"A succinct text-book of chemical analysis which may and should be used with any good text-book of principles."

Pott 8vo.   3s.

Questions on Chemistry.
A Series of Problems and Exercises in Inorganic and Organic Chemistry. By FRANCIS JONES, F.R.S.E., F.C.S., Chemistry Master in the Grammar School, Manchester.

THE EDUCATIONAL TIMES.—"Certainly this book of questions cannot fail to be useful in fixing the most important facts of chemistry in the student's mind."

THE SCOTSMAN.—"They present a very exhaustive statement of the elementary problems of both organic and inorganic chemistry, and the pupil who works the solutions of them all may with confidence consider that he possesses a thorough acquaintance with the rudiments of science."

4to.   7s. net.

Qualitative Analysis Tables and the Reactions
of Certain Organic Substances. By E. A. LETTS, D.Sc., PH.D., F.R.S.E., F.C.S., etc., Professor of Chemistry in Queen's College, Belfast.

THE BRITISH MEDICAL JOURNAL.—"The work before us is certainly one of the best of its kind, and is free from objection if intelligently used by both teacher and student. The tables for the reactions of the alkaloids and of normal and morbid urine, as well as those for the examination of blood, urinary calculi, etc., will be found extremely useful to medical men. A special feature in the book is a very carefully prepared table of the solubilities of the salts of the various metals in the principal solvents."

THE LANCET.—"It is evidently the outcome of experience as to the requirements of general students, and, above all, it has not been written apparently with the view of rushing the student through any particular examination. The solubility table (IX.), though somewhat complex, will be useful in conveying to the freshman information on the salts which are, and which are not, soluble. . . . . The articles on the reactions and properties of some of the more important organic substances, with tables for their detection, are good, and should be of value to the student of medicine. The pages are boldly printed in clear type and the book is commendably free from errors."

Crown 8vo. 6s.

An Introduction to the Science and Practice
of Qualitative Chemical Analysis, Inorganic. By CHAPMAN JONES, F.I.C., F.C.S. (London and Berlin), etc., the Senior Demonstrator of Practical Chemistry at the Royal College of Science, London; Assistant Examiner in Chemistry to the Department of Science and Art, etc.

Crown 8vo. 7s. 6d.

An Elementary Text-Book of Chemistry.
By WILLIAM G. MIXTER, Professor of Chemistry in the Sheffield Scientific School of Yale University. Second and Revised Edition.

THE PRACTITIONER.—"This little book will thus aid him not only to pass his professional examination in practical chemistry more easily, but will give him such an insight into the subject as will enable him readily to extend his knowledge of it, should time and inclination permit."

NATURE.—"The book is a useful compendium of the principal properties of not only the more important substances but also of many of the compounds of the rarer elements."

Globe 8vo. 2s. 6d.

Introduction to the Study of Chemistry.
By W. H. PERKIN, jun., Professor of Chemistry in the Owens College, Manchester, and BEVAN LEAN, D.Sc., B.A. (Lond.), Assistant Lecturer and Demonstrator and late Berkely Fellow of the Owens College, Manchester. Illustrated.

THE GUARDIAN.—"This book lays great stress on early quantitative work by all the students themselves. Undoubtedly such work is more easy to secure in University colleges and polytechnics than in the average secondary schools. But wherever the necessary bench accommodation can be obtained, this is the book which in our judgment, is—since the days of Galloway's *First and Second Steps in Chemistry*—best qualified to stimulate the interest and evolve the mettle of the student."

THE EDUCATIONAL TIMES.—"We welcome this book as containing the best treatment of chemistry as an educational subject that we have yet seen. . . . We heartily recommend this book to all schoolmasters and science teachers who desire to utilize experimental chemistry as a really educational subject."

THE ATHENÆUM.—"The authors are to be most sincerely congratulated both on their effort to produce a good introduction to the study of natural philosophy and on their accomplishment; the method and the matter are alike excellent, and the historical notes and anecdotes occasionally quoted add a living and human interest to the subject which will serve to attract many a young student. We can most cordially recommend this little book and wish it every success."

Pott 8vo. 2s. 6d.

Experimental Proofs of Chemical Theory for
Beginners. By WILLIAM RAMSAY, PH.D., Professor of Chemistry in University College, London.

THE SATURDAY REVIEW.—"A thoroughly useful guide in a most important department of chemistry. . . . The field covered by the manual includes such subjects as measurement of temperatures, pressure and weight, relations of the volume of gases to temperature and pressure, air, water, atoms and molecules, specific heat, replacement, periodic law, and so on ; so that it treats practically of the subjects that lie at the very foundation of the science. The book ought to find its way into chemical laboratories."

THE SCOTSMAN.—" It is, in fact, an elementary manual of laboratory work. It describes in detail the experiments by which the fundamental principles and laws of chemistry are demonstrated, and contains clear directions for the construction of the requisite apparatus."

Crown 8vo. 6s. 6d.

An Introduction to the Study of Chemistry:
Inorganic Chemistry. By IRA REMSEN, Professor of Chemistry in the John Hopkins University.

THE SATURDAY REVIEW.—" It professes to be more systematic than the ordinary elementary manuals, and is certainly of a thoroughly practical character, especially in regard to laboratory instruction, which is almost a crucial test. Mr. Remsen's style and reasoning are clear, plain, and free of pedantry."

THE ATHENÆUM.—" The introductory manual which has been written by Professor Remsen, of Baltimore, may be pronounced without hesitation to be an excellent little book—excellent alike in aim and method. Its aim is to treat the subject scientifically, and its method is to do this by a succession of experiments systematically arranged ; the observation of phenomena being put in the front, while all theory is kept well in the background. It is evident that Professor Remsen holds very sound views on the matter of scientific education. . . . His book will do much towards placing the teaching of elementary chemistry on a more rational basis."

Crown 8vo. 6s. 6d.

An Introduction to the Study of the Compounds
of Carbon: or Organic Chemistry. By IRA REMSEN, Professor of Chemistry in the John Hopkins University.

THE LANCET.—" A clear account of the principles, including also the preparation of important organic substances."

NATURE.—" In no other elementary book in the English language will the student find so many admirably chosen examples of the formation of structural formulæ. The important facts are noted ; then the inference is drawn ; then the hypothesis is ventured upon ; analogous facts are recalled ; the hypothesis is strengthened or weakened ; suggestions are made ; experiments are conducted ; and all is finally summarised in the formula. . . . The author of this book deserves the thanks of all chemical teachers who have tried to teach organic chemistry to beginners, for the clear and short directions which he gives for preparing the more important compounds of carbon. The book may well be used as a laboratory guide, no less than as an introduction to the science of organic chemistry."

THE NATION (New York).—" Of Dr. Remsen's qualification for the task he has undertaken, it is needless here to speak. He is widely known, not only as an eminent chemist, but also as a clear and careful writer on chemical themes—a practical educator, intent on teaching students to *think*. He makes no apology for his book, nor needs to, for it will be welcomed by all who are anxious to see chemical teaching lifted out of the rut into which it has for the most part fallen. His book abounds in well-selected, explicitly described experiments, set in a text which is always clear and attractive, and often very suggestive."

Foolscap 8vo. 2s. 6d.

The Elements of Chemistry.

A Text-book for beginners. By IRA REMSEN, Professor of Chemistry in the John Hopkins University.

THE JOURNAL OF EDUCATION.—" This is evidently the work of a practical teacher, and the subjects of the first thirteen chapters are most judiciously arranged and thoroughly well worked out..... A good series of questions is appended."

THE LITERARY WORLD.—"About half the book is taken up with a description of chemical change, air, water, the more common gases, and their properties. We next get a short description of the common metals and their salts, with the relation existing between them. Then follow two chapters on the more familiar compounds of carbon—as soap, alcohol, vinegar, sugar, starch, &c..... This work is very well suited as an introductory text-book for schools."

Foolscap 8vo. 4s. 6d.

Lessons in Elementary Chemistry.

By Sir H. E. ROSCOE, F.R.S. With Illustrations. Sixth Edition.

THE CHEMICAL NEWS.—" It has no rival in its own field, and it can scarcely fail to take its place as the text-book at all schools where chemistry is now studied..... Well fitted for the junior chemical classes at the universities, colleges and medical schools."

THE SPECTATOR.—" This is a small, compact, carefully elaborated and well-arranged manual of elementary chemistry in all its branches. For the correctness and accuracy of its contents the name of the writer is a sufficient guarantee, holding as he does no mean rank amongst English chemists, and having given close attention not to one only, but to many branches of his subject. Small as it is, as a standard general text-book it deserves to take a leading place.... A carefully prepared series of questions added in an appendix is a most useful addition to the book."

Globe 8vo. Price 2s. 6d.

Inorganic Chemistry for Beginners.

By Sir H. E. ROSCOE, F.R.S. Assisted by JOSEPH LUNT, B.Sc. (Victoria).

NATURE.—"In every respect the book is a good one, and contains the kind of knowledge that should be imparted to all beginners of science."

THE ATHENÆUM.—" We heartily commend the book as an introduction to the elementary principles of chemistry."

THE SPEAKER.—" This is the kind of text-book that should be multiplied..... At the end of each of the twenty-one lessons will be found a concise recapitulation of points treated, and exercises testing knowledge. The little book is full of matter and full of teaching power."

Globe 8vo. 2s. 6d.

Practical Inorganic Chemistry for Advanced

Students. By CHAPMAN JONES, F.I.C., F.C.S. (London and Berlin), etc., the Senior Demonstrator of Chemistry in the Royal College of Science, London; Assistant Examiner in Chemistry to the Department of Science and Art.

Globe 8vo. 2s. 6d.

Practical Inorganic Chemistry. By G. S. Turpin, M.A., D.Sc., Head Master of the Intermediate and Technical School, Swansea.

NATURE.—"It encourages thought, creates interest in chemistry and furnishes the kind of knowledge most likely to prove of advantage in after years. Not only in organised science schools, but in every school where chemistry is taught, the course described in this book could be profitably introduced."

THE EDUCATIONAL NEWS.—"To all students of inorganic chemistry this book may safely be recommended, but to those who to make themselves proficient in the subject desire practical work, it will be a decided acquisition. In all respects it is a very suitable manual."

THE SPEAKER.—"Its thorough and workmanlike character is sure to recommend this book to the managers of science schools."

Globe 8vo. 2s. 6d.

Lessons in Organic Chemistry. Part I. Elementary. By G. S. Turpin, M.A., D.Sc., Head Master of the Intermediate and Technical School, Swansea. Part II. [*In preparation.*

THE SATURDAY REVIEW.—"The treatment is sound and the arrangement clear."

THE JOURNAL OF EDUCATION.—"The author writes in a clear, concise style. The sentences are short and to the point, and can be easily followed. It will be found a very useful book to place in the hands of pupils. The illustrations, thirty-three in number, are very distinct. A few good questions are appended to each chapter. The book has been very carefully written and arranged, and can be well recommended for class use."

CHEMISTRY.

ADVANCED.

8vo. 15s. net.

A Dictionary of Chemical Solubilities, Inorganic. By Arthur Messinger Comey, Ph.D., formerly Professor of Chemistry, Tufts College.

THE ACADEMY.—"An invaluable addition to the library of the chemist."

THE WESTMINSTER REVIEW.—"No chemist can consider his library complete without a copy of this useful work."

KNOWLEDGE.—"A chemical library will now not be complete without this monumental work . . . . and every practical chemist should endeavour to obtain it."

Crown 8vo.   7s. 6d.

A System of Volumetric Analysis. By Dr. EMIL FLEISCHER.
Translated with notes and additions from the Second German Edition by M. M. PATTISON MUIR, F.R.S.E., Assistant Lecturer on Chemistry, the Owens College, Manchester. With Illustrations.

> *NATURE.*—"Mr. Muir has done a very considerable service to his chemical brethren by his translation of Dr. Fleischer's work, ... the first attempt at a rational system of volumetric analysis which has been offered to us in this country. .... Part II. .... constitutes a most valuable addition to the art of chemical analysis. .... A book which is really a solid contribution to chemical literature."

Crown 8vo.   8s. 6d.

The Practical Methods of Organic Chemistry.
By LUDWIG GATTERMANN, PH.D., A. O. Professor in the University of Heidelberg. Authorised Translation by WILLIAM B. SHOBER, PH.D., Instructor in Organic Chemistry in the Lehigh University. With numerous Illustrations.

> *NATURE.*—"It would be hard, indeed, to express anything but a very favourable opinion of Prof. Gattermann's excellent book as an introduction to practical organic chemistry: a student who reads it carefully will save himself labour, time, and material, and will avoid many of the usual mistakes and accidents ; at the same time he will gain a sound practical knowledge which will help him to commence research with a good prospect of success. Dr. Shober's translation is very readable."

Crown 8vo.   7s. 6d.

Methods of Gas Analysis. By Dr. WALTHER HEMPEL,
Professor of Chemistry, the Dresden Polytechnicum. Translated from the Second German Edition by L. M. DENNIS, Assistant Professor of Analytical Chemistry in Cornell University.

> *THE CHEMICAL NEWS.*—"If we pronounce this work not merely good and useful, but even admirable, we shall not be guilty of exaggeration."

> *THE ACADEMY.*—"The volume constitutes a general text-book of select processes. It is written in an interesting and easy style, and shows everywhere the practised hand of a master familiar with all the ins and outs of the operations which he describes."

> *THE NATION.*—"The work is very complete and forms an extremely useful and valuable addition to the resources of the analyst."

Crown 8vo.   8s. 6d.

A Laboratory Manual of Organic Chemistry.
A Compendium of Laboratory Methods for the use of Chemists, Physicians and Pharmacists. By Dr. LASSAR-COHN, Professor of Chemistry in the University of Königsberg. Trans-

lated, with the author's sanction, from the Second German Edition by ALEXANDER SMITH, B.Sc., PH.D., Assistant Professor of General Chemistry in the University of Chicago.

THE CHEMICAL TRADE JOURNAL.—"This book, which is admirably got up, is one of the most useful of recent additions to the English literature of organic chemistry, and will be found invaluable to every practical student or chemist engaged in organic work."

THE SATURDAY REVIEW.—"Altogether this is an extremely satisfactory book in its minor details as in its essential plan, and one which we are glad to recommend unreservedly to the laboratory student of organic chemistry."

Crown 8vo. 6s.

The Elements of Electrochemistry. By MAX LE BLANC, Professor of Chemistry in the University of Leipzig. Translated by W. R. WHITNEY, Instructor of Chemistry in the Massachusetts Institute of Technology of Boston, U.S.A.

THE EDUCATIONAL REVIEW.—"The book is well up to date, and sound and clear in its method of exposition, and will be of great service in introducing those who have mastered the ordinary elementary text-books, to the methods of higher research. It is based on experimental methods throughout, and its mathematics should not be too great a tax on the ordinary science student. The typography and general style of production is admirable."

THE JOURNAL OF EDUCATION.—"Students of the higher branches of physical chemistry, of which one of the most interesting and progressive is electro-chemistry, are under an obligation to Mr. Whitney for translating this useful book. The translator has produced an accurate and yet readable translation . . . . It is a useful book, and one which at present has no rival in the language."

Medium 8vo. 17s. net.

Analytical Chemistry. By A. MENSCHUTKIN, Professor in the University of St. Petersburg. Translated from the third German Edition under the supervision of the author by JAMES LOCKE.

THE EDUCATIONAL REVIEW.—"This book will at once take rank as one of the foremost standard works as yet published on qualitative and quantitative analysis. . . . . The translation has been prepared under the author's own supervision, and is admirably done."

NATURE.—"In the work before us, the subject is presented in such a clear and original manner, that it can hardly fail to become as popular in this country as in Germany. . . . . The descriptions are concise and the methods well chosen."

8vo. 15s. net.

A History of Chemistry from the Earliest Times to the Present Day. Being also an Introduction to the Study of the Science. By ERNST VON MEYER, PH.D., Professor of

Chemistry in the Technical High School, Dresden. Translated, with the Author's sanction, by GEORGE M'GOWAN, PH.D., Demonstrator in Chemistry, University College of North Wales, Bangor. Second Edition.

THE SPEAKER.—" The conclusions of Prof. v. Meyer are hardly likely to be disputed, for they are obviously the outcome of wide knowledge, independent research, and sound judgment.... Dr. M'Gowan has rendered the original text of the book into excellent English, and the marks of more than ordinary care prevail on almost every page."

THE PALL MALL GAZETTE.—" This is a most interesting work, and will be read by the thoughtful student of chemistry with both pleasure and profit.... The translator has done his part in such a way as leaves nothing to be desired."

8vo. 12s. 6d.

The Elements of Thermal Chemistry. By M. M. PATTISON MUIR, M.A., F.R.S.E., Fellow and Prælector in Chemistry of Gonville and Caius College, Cambridge, assisted by DAVID MUIR WILSON. With a plate.

THE ATHENÆUM.—" The general impression which the book gives is that it is the work of a candid and intelligent inquirer who finds considerable difficulties in the attainment of truth or of logical consistency. It contains a large body of well-arranged information, not easily accessible elsewhere to the ordinary student, and will doubtless be extensively used as a text-book."

THE SATURDAY REVIEW.—"An able and trustworthy attempt to systematize recent investigations and theories regarding the distribution of energy in chemical phenomena..... After illustrating the methods of inquiry, the book shows how thermal methods are applied to chemical questions and also to some which are only partially chemical, concluding with a discussion of the 'chemical interpretation to be given to thermal data in the light of the prevalent theories' regarding matter and energy."

8vo. 15s. net.

Theoretical Chemistry from the Standpoint of Avogadro's Rule and Thermo-Dynamics. By Professor WALTER NERNST, PH.D., of the University of Göttingen. Translated by Professor CHARLES SKEELE PALMER, PH.D., of the University of Colorado. With 26 woodcuts and 2 appendices.

THE JOURNAL OF EDUCATION.—"It is the best work in the language on advanced physical chemistry."

THE ACADEMY.—" Dr. Nernst's treatise is certain to prove not only a source of sound instruction to the student, but a powerful stimulus to the investigator."

THE OXFORD MAGAZINE.—"The book ... is indispensable to every advanced student."

8vo. 10s. net.

Outlines of General Chemistry. By WILHELM OSTWALD, Professor of Chemistry in the University of Leipzig.

Translated, with the Author's sanction, by JAMES WALKER, D.Sc., PH.D., Professor of Chemistry in University College, Dundee.

NATURE.—"It furnishes a well-conceived syllabus of the subject-matter of general and physical chemistry. . . . . The general conception of the book is admirable; it contains much that is new; to the advanced reader it will be refreshing after the time-honoured methods of the ordinary text-books . . . . . The work from its very title, apart even from the reputation of the author, will no doubt appeal to a large class of readers; as an English text-book of chemistry it is unique."

THE SPECTATOR.—"Just the sort of treatment which the general student is thankful for. . . . . The style is clear and expresses the matter well."

8vo. 7s. net.

Manual of Physico-Chemical Measurements.

By WILHELM OSTWALD, Professor of Chemistry in the University of Leipzig. Translated, with the Author's sanction, by JAMES WALKER, D.Sc., PH.D., Professor of Chemistry in University College, Dundee. Illustrated.

THE ATHENÆUM.—"In the compass of about 250 octavo pages a vast amount of well-digested information is given, introducing the student to the best modern methods. There are numerous illustrations well-suited for explaining the text, and the style, though brief and note-like, is clear as far as it goes. The manual will be indispensable as a book of reference in every physical laboratory."

THE ACADEMY.—"In the clearness of its 188 figures, in the ingenuity of many of its minute contrivances, in its judicious estimations of the relative accuracy of methods, and above all in its sound descriptions and discussions of the bases of calculation, this manual presents features of unusual excellence."

Crown 8vo. 5s. net.

The Scientific Basis of Analytical Chemistry.

Treated in an Elementary Manner. By WILHELM OSTWALD, Professor of Chemistry in the University of Leipzig. Translated, with the author's sanction, by GEORGE M'GOWAN, PH.D.

THE UNIVERSITY CORRESPONDENT.—"As an exposition of the principles of qualitative and quantitative analysis it is wholly admirable, and must be read by every student of scientific chemistry. The translation has been so well done that not a sentence reads as though it were a translation from the German."

THE CHEMICAL TRADE JOURNAL.—"The book appeals more directly to students, but at the same time it is as well to ask who—even the most learned—is not yet still a student of chemistry? A perusal will never prove waste of time, while to the majority it is as necessary as it is helpful."

Medium 8vo. 16s.

Inorganic Chemistry.

By IRA REMSEN, Professor of Chemistry in the John Hopkins University.

INDUSTRIES.—"Professor Remsen has conscientiously performed the task which he had in view."

THE JOURNAL OF EDUCATION.—"A most interesting volume is the result, and one which cannot fail to be suggestive to any thoughtful student. . . . We recommend this volume to anyone wishing to get a sound knowledge of Inorganic Chemistry; it has not the slightest trace of cram system, but fully sustains the author's reputation as a clear, thorough, interesting and suggestive teacher of the science. Descriptions of experiments, illustrative of the text, are collected and placed at the end of the book."

In 3 Vols. 8vo.

A Treatise on Chemistry. By Sir H. E. Roscoe, F.R.S., and C. Schorlemmer, F.R.S.

Vol. I. The Non-Metallic Elements. 21s. New Edition, completely revised by Sir H. E. Roscoe, assisted by Drs. H. E. Colman and A. Harden. With Illustrations and a portrait of Dalton engraved by C. H. Jeens.

Vol. II. Metals. 31s. 6d.

Vol. III. The Chemistry of the Hydrocarbons and their derivatives, or Organic Chemistry. Part I. 21s. Part II. 21s. Part III. 18s. Part IV. 21s. Part V. 18s. Part VI. 21s.

Of Volume I.

NATURE says:—"The student has presented to him in this volume of the 'Treatise' the material that is needed for acquiring a real knowledge of the chemistry of the non-metallic elements."

THE CAMBRIDGE REVIEW says:—"No further commendation will be required than to say that, although completely revised and rewritten, it carries out the aims put forward in the first preface seventeen years ago."

Of Volume II.

THE LANCET says:—"This most valuable manual.... It would be difficult to praise the work too highly. All the merits which we noticed in the first volume are conspicuous in the second. The arrangement is clear and scientific; the facts gained by modern research are fairly represented and judiciously selected; and the style throughout is singularly lucid."

Of Volume III.

THE EDINBURGH MEDICAL JOURNAL says:—"The book is a most valuable contribution to Science, and worthy of the high reputation of its authors."

NATURE says —"The information is very full. The interesting theoretical and historical discussions are continued throughout the volume, and impart to it a character of 'readableness' rather unusual in a work of this nature. Finally, the student of Technology will find the various manufacturing processes treated of in some detail."

Medium 8vo. 21s.

Spectrum Analysis. Six Lectures delivered in 1868 before the Society of Apothecaries of London. By Sir H. E. Roscoe, F.R.S. With appendices, coloured plates, and illustrations. Fourth Edition, revised and considerably enlarged by the Author and Professor Arthur Schuster.

THE BRITISH MEDICAL JOURNAL.—"There is, we believe, no modern discovery in physical science which has been so remarkable, or has made such rapid progress as Spectrum Analysis, and a more complete, and at the same time more simple and intelligible, account of the subject could scarcely be given, than is contained in the book before us."

THE WESTMINSTER REVIEW.—"The lectures themselves furnish a most admirable elementary treatise on the subject, whilst by the insertion in the appendix of extracts from the most important published memoirs, the author has rendered it equally valuable as a text-book for advanced students."

Crown 8vo. 5s.

The Rise and Development of Organic Chemistry.
By CARL SCHORLEMMER, LL.D., F.R.S., late Professor of Organic Chemistry in the Owens College, Manchester. Edited by ARTHUR SMITHELLS, B.Sc., Professor of Chemistry in the Yorkshire College, Leeds.

THE BRITISH MEDICAL JOURNAL.—"We would most heartily commend this volume to members of the medical profession. A more living and suggestive account of one of the scientific aspects of some of our daily problems could scarcely be imagined."

THE SPEAKER.—"It discloses at once the breadth and depth of Schorlemmer's knowledge, and his ardent and disinterested love of what he was wont to term, with the pride of a disciple, 'Our Science.' A fine photogravure portrait of the authors forms the frontispiece of the volume."

Foolscap 8vo. 2s.

A Series of Chemical Problems.
With Key. For use in colleges and schools. By T. E. THORPE, PH.D., B.Sc. Vic., Sc.D. Dubl., F.R.S., Professor of Chemistry in the Royal College of South Kensington. Revised and enlarged by W. TATE, Assoc. N.S.S. (Honours), F.C.S. With a preface by Sir H. E. ROSCOE, B.A., PH.D., F.R.S., M.D. New Edition.

Crown 8vo. 8s. 6d. net.

Essays in Historical Chemistry.
By T. E. THORPE, PH.D., B.Sc. Vic., Sc.D. Dubl., F.R.S., Professor of Chemistry in the Royal College of South Kensington.

THE ATHENÆUM.—"The author has in this collection of essays told something of the work and the lives of some men of science, and has done this in a pleasing and sympathetic manner; for this his readers will cordially thank him. No follower of the science of chemistry can well be without this little volume, and a large number of others who may have little knowledge of chemistry will peruse it with great pleasure and no little profit."

THE JOURNAL OF EDUCATION.—"The book will not only appeal to the student, who will find the references to published papers of great service to him, but the general reader will find it a most entertaining volume of great literary excellence."

Crown 8vo. 6s.

A History of Chemical Theory from the Age of Lavoisier to the Present Time.
By AD. WURTZ, Membre de L'Institut, Académie des Sciences. Translated and edited by HENRY WATTS, B.A., F.R.S.

THE CHEMICAL NEWS.—"Chemical literature in England owes a great debt to Mr. Watts, and we sincerely hope that his edition of M. Wurtz's *discours préliminaire* will aid in reviving a taste for a neglected but most important branch of literature."

THE ATHENÆUM.—"Mr. Watts has executed the translation admirably; and by his elucidation of the few discourses having important bearing on the subject, which were passed over by M. Wurtz, he has greatly added to the value of the volume."

PHYSICS.

ELEMENTARY.

Foolscap 8vo.   4s. 6d.

Physics for Students of Medicine. By ALFRED DANIELL, M.A., LL.B., D.Sc., F.R.S.E., of the Inner Temple, Barrister-at-Law, Member of the Faculty of Advocates in Scotland, formerly Lecturer on Physics in the School of Medicine, Edinburgh.

THE LANCET.—"This is a recently published work by Professor Daniell, whose experience of the requirements of medical students in regard to physics is well known. It will be found an excellent companion to lecture work, and is suitable for students preparing for the elementary physics examination of the Conjoint Board."

NATURE.—"We commend Dr. Daniell's volume to teachers of physics generally, believing that they will find it worthy of adoption. The contents include chapters on units of measurement, motion of bodies, friction, matter, sound, heat, ether-waves, and electricity. All these subjects are treated as thoroughly as is possible in a book of this character."

Crown 8vo.   5s.

Practical Lessons in Physical Measurement. By ALFRED EARL, M.A., late Scholar of Christ's College, Cambridge, of the Inner Temple, Barrister-at-Law, Senior Science Master at Tonbridge School.

THE EDUCATIONAL TIMES.—"We recommend it unreservedly to the attention of every head-master. . . . We consider that Mr. Earl has done something more than produce a useful and honest book: he has distinctly advanced the cause of genuine science teaching in the country by giving us a practical demonstration of what genuine science teaching may be. From that point of view it is one of the most important contributions to educational literature that have appeared this year."

THE OXFORD MAGAZINE.—"The student who uses this book will be educated as well as trained. He will learn how to measure and how to understand the meaning of his results. . . . We think that the present work should receive a hearty welcome."

Foolscap 8vo.   3s.

Elements of Physics. By C. E. FESSENDEN, Principal of the Collegiate Institute, Peterboro', Ontario.

NATURE.—"Forms an excellent introduction to a more extended study of physical science . . . For class instruction in schools . . . it is admirably well adapted. . . . The style is concise, but clear and accurate, and as the book has not been written with the view of preparing the student for any special examination, it is refreshingly free from any tendency towards cram."

Foolscap 4to.   2s. 6d.

An Exercise Book of Elementary Practical
Physics for Organised Science Schools under the Department of Science and Art Evening Continuation Schools, and Elementary Day Schools. Arranged according to the Head Masters' Association's Syllabus of Practical Physics. By RICHARD A. GREGORY, F.R.A.S.

THE SATURDAY REVIEW.—"In no book that we know of is this method quite so admirably set forth as in this admirable book of exercises that Mr. Gregory has arranged. . . . Here and there he shows himself not only a scientific teacher, but a remarkably ingenious one."

THE EDUCATIONAL REVIEW.—"Will be found extremely useful to students in organised science schools under the Department of Science and Art, in evening continuation schools, and in the highest standards in elementary schools."

Globe 8vo.   2s. 6d.

An Introduction to Practical Physics for use in
Schools. By D. RINTOUL, M.A., Assistant Master at Clifton College, and sometime Fellow of Corpus Christi College, Cambridge.

Foolscap 8vo.   3s. 6d.

Examples in Physics.
Containing over 1000 problems with answers and numerous solved examples. Suitable for candidates preparing for the Intermediate, Science Preliminary, Scientific and other Examinations of the University of London. By D. E. JONES, B.Sc. (Lond.), Director of Technical Instruction to the Staffordshire County Council, late Professor of Physics in the University College of Wales, Aberystwyth. Second Edition.

NATURE.—"In the chapters on dynamics, hydrostatics, heat, light, sound, electricity and magnetism—chapters which consist essentially of examples—there are clearly-written paragraphs explaining those points that do not generally seem to be grasped by students. . . . The general arrangement of the book is particularly happy; it is clearly the work of a teacher whose object is to increase the real knowledge of his students and not merely to drive them through the ordeal of an examination."

THE ATHENÆUM.—"Gives a clear exposition of physical units, some hints on approximate calculations, and an excellent collection of examples on all branches of physics. . . . Brief explanations are given where necessary and some useful tables are introduced."

Globe 8vo.   2s. 6d.

Elementary Lessons in Heat, Light and Sound.
By D. E. JONES, B.Sc. (Lond.), Director of Technical In-

struction to the Staffordshire County Council, late Professor of Physics in the University College of Wales, Aberystwyth. *Heat*, separately. Sewed. 1s.

THE SCHOOLMASTER.—"May be confidently recommended for giving sound elementary knowledge of these branches of physical science. The statements are concise, the definitions clear and accurate, and the descriptions, both verbal and pictorial, are well chosen."

THE ATHENÆUM.—"This is an excellent work of 282 pages, conscientiously and carefully constructed, with clear explanations and well-executed diagrams. We hope its success may be proportionate to its merits."

Foolscap 8vo. 3s. 6d.

Lessons in Heat and Light. By D. E. JONES, B.Sc.

(Lond.), Director of Technical Instruction to the Staffordshire County Council, late Professor of Physics in the University College of Wales, Aberystwyth. Illustrated.

NATURE.—"As an introduction to the study of experimental physics, the book cannot fail to be of great value. The principles of the subjects are very clearly stated, and the experiments from which they have been deduced are fully described. Most of the experiments may be easily performed by students, the instructions being sufficiently clear to guarantee success."

THE ELECTRICIAN.—"We think every elementary student of physics should possess a copy of this work."

Pott 8vo. 2s.

Questions and Examples on Elementary Experi-

mental Physics, Sound, Light, Heat, Electricity and Magnetism. By BENJAMIN LOEWY, F.R.A.S., etc., Examiner in Experimental Physics at the College of Preceptors, London.

THE EDUCATIONAL TIMES.—"The practical and suggestive character of Mr. Loewy's examination questions has often been recognized by both teachers and examiners, but now that the questions are classified and arranged in logical order these qualities are still more evident, and the collected series forms a very useful guide for students of Elementary Physics."

THE CAMBRIDGE REVIEW.—"This little book supplies a distinct want. Though primarily intended for beginners, it is yet well suited for testing much higher classes of students, for example, candidates for Natural Science Scholarships or the first M.B. The capability of answering all these questions satisfactorily would imply a pretty complete knowledge of the groundwork of the experimental theories of Sound, Light, Heat and Electricity."

Foolscap 8vo. 2s. 6d.

Numerical Tables and Constants in Elementary

Science. By SYDNEY LUPTON, M.A., Assistant Master in Harrow School. With a map and table.

NATURE.—"Mr. Sydney Lupton, Assistant Master of Harrow School, has recently compiled and published some numerical tables and constants in elementary science which we can fairly recommend to our readers. . . . . The book deals with numbers and measures, heat, light, sound, electricity, chemistry and physiography; the latter division being wide enough to include tables of logarithms."

Crown 8vo. 2s. 6d.

Light. A series of simple, entertaining and inexpensive experiments in the phenomena of light. For the use of students of every age. By ALFRED M. MAYER and CHARLES BARNARD. With Illustrations.

> *IRON.*—"It is a masterpiece both in method and expression. . . . . In the book before us, in the compass of ninety pages, we have a methodical work specially designed for the young and for the teacher who, though lacking apparatus, wishes his pupils to become experimenters, strict reasoners and exact observers. In accordance with this plan all the experiments described are new, the materials employed are purposely of the cheapest description; and while the apparatus may be purchased or made at small cost, it is at the same time suitable for regular daily use at home or in school, and such as with moderate care will last for years. . . . . Messrs. Macmillan and Co. have reason to be gratified with this contribution to popular education."
>
> *THE SPECTATOR.*—"The experiments are well devised, easy to execute, the apparatus required is not costly, and the knowledge to be derived from this simple course is exceedingly valuable and a good foundation for a more extended study of the subject."

Crown 8vo. 7s. 6d. net.

The Outlines of Physics. An Elementary Text-Book. By EDWARD L. NICHOLS, Professor of Physics in Cornell University.

> *THE SPEAKER.*—"A good elementary text-book of Physics for use in connection with laboratory work. . . . The volume is not to be regarded simply as a laboratory manual, but is intended to fill the place of a text-book and laboratory guide combined, in cases where the amount of time allotted to physics does not make it advisable to use both a text-book and a laboratory manual."
>
> *THE GLASGOW HERALD*—"A really excellent publication. . . . The style of treatment generally it would be hard to improve upon. The illustrations are very numerous and excellent."

8vo. 12s. 6d. net.

A Laboratory Manual of Physics and Applied Electricity. Arranged and Edited by EDWARD L. NICHOLS, Professor of Physics in Cornell University. In Two Volumes. Illustrated.

Vol. I. Junior Course in General Physics. By ERNEST MERRITT and FREDERICK J. ROGERS.
(For Vol. II, see under Physics, Advanced.)

> *THE SPEAKER.*—"It seems to us that this, in the scientific sense, is an admirable book, and one which is likely to fill for a considerable time to come a hitherto distinct lack in the literature of education."
>
> *THE NATION.*—"Admirably adapted to supply the needs of the modern physical laboratory."
>
> *THE OXFORD MAGAZINE.*—"We know no text-book in English which covers the same ground, and can cordially recommend it to the notice of teachers and students."

In 3 vols. 8vo. 6s. net each.

The Elements of Physics. A College Text-Book. By EDWARD L. NICHOLS and WILLIAM S. FRANKLIN.

    Vol. I. Mechanics and Heat.
    Vol. II. Electricity and Magnetism.
    Vol. III. Light and Sound.

Of Volume I.

THE JOURNAL OF EDUCATION says:—"It is thoroughly scientific and concise. Altogether a useful text-book and one which we feel no hesitation in recommending to students working for University science degrees."

Of Volume II.

THE GUARDIAN says:—"Well illustrated and beautifully printed, the writers have admirably succeeded in what they aimed at—a college text-book of modern electrical science, concise but precise, and written from the physical standpoint with remarkable balance of treatment."

Of Volume III.

THE ACADEMY says:—"Should find a place among the books used in our university colleges and institutions of the same rank."

Foolscap 8vo. 4s. 6d.

Lessons in Elementary Physics. By BALFOUR STEWART, M.A., LL.D., F.R.S., late Langworthy Professor of Physics, Owens College, Manchester, and Professor of the Victoria University. With a coloured spectrum. New and enlarged edition.

THE LITERARY WORLD.—"We must again commend this book as a most interesting and trustworthy guide to students of physics."

THE EDUCATIONAL TIMES.—"The revision has been completed by Mr. W. Haldane Gee. The chief changes introduced have been in the chapters on Electricity, which have been re-arranged and new figures and additional matter included."

THE GUARDIAN.—"Every endeavour has been made to keep the work up to date, especially under the head of electricity and magnetism, so that it is still the best fitted for its purpose."

Pott 8vo. 2s.

Questions on Stewart's Lessons in Elementary Physics. By T. H. CORE.

THE WESTMINSTER REVIEW.—"The questions appear to be excellent, and there can be no doubt but that they may be useful alike to students and teachers."

In 3 vols. Crown 8vo.

Lessons in Elementary Practical Physics. By BALFOUR STEWART, LL.D., F.R.S., late Langworthy Professor of Physics, Owens College, Manchester, and Professor

of the Victoria University, and W. W. HALDANE GEE, Demonstrator and Assistant Lecturer in Physics, the Owens College.

Vol. I. General Physical Processes. 6s.
Vol. II. Electricity and Magnetism. 7s. 6d.
Vol. III. Part I. Practical Acoustics. By C. L. BARNES, M.A., F.C.S. 4s. 6d.
[Part II., Heat; Part III., Optics, *in preparation*.

Of Volume I.

THE ATHENÆUM says:—"Professor Balfour Stewart having been formerly for many years Director of Kew Observatory, is eminently qualified to speak with authority as to the methods in actual use for making measurements with all possible accuracy, and the volume to which his name is attached gives very full directions for using various instruments for the measurements of lengths, thicknesses, angles, masses, areas, volumes, densities, and other data belonging to mechanics and hydrostatics. . . . It is clearly and judiciously written, and is a great advance upon anything of the same kind previously existing in the English language."

NATURE says:—"In this the first volume of what will evidently be an elaborate work on practical physics, the authors have treated of general physical processes only, *i.e.* of the methods employed in the laboratory for the exact determination or measurement of the geometrical and mechanical properties of bodies. It is impossible to overestimate the importance of these fundamental measures, for upon them depends the accuracy of almost all physical work. That this is the view of the authors is made evident by their having devoted nearly the whole of the first volume out of a promised three to matters purely geometrical and mechanical. Throughout the volume the most minute attention to details is apparent, so much so that it is likely to weary those who read it only; but those who use it to guide them in making the measures given will certainly benefit by the completeness with which each subject is treated."

Of Volume II.

THE JOURNAL OF EDUCATION says:—"This volume of the practical work of the two Owens College teachers is a delight. It shows us how completely the actual manipulation of apparatus, and calculations based upon observation, are replacing the mere reading-up of tangent galvanometers and the working out of entirely theoretical examples. Manipulation, observation, recorded in terms of mathematics—these are the new methods of modern physics. Not that in this work there is no touching upon theoretical points. That is done to a considerable extent, but always in connection with practical work that makes the abstract theories concrete realities to the student. The apparatus needed is simple; method and instruments are generally typical; the mathematics required are not too abstruse: constant reference is given to more advanced works. Not a book to be read straight through, this work of Messrs. Stewart and Gee is one to be worked steadily through by any student who intends to acquire a real knowledge of the principles of Electricity and Magnetism."

INDUSTRIES says:—"The second volume of this admirable work—succinctly and accurately described by its brief title—fully bears out the promise of the first, which was published about two years ago. . . . An admirable and most useful book."

THE WESTMINSTER REVIEW says:—"The book is essentially an exposition of those parts of electrical philosophy with which the student who would become an electrician must necessarily be acquainted. It is well illustrated, clearly arranged, well printed, and is likely to be a valuable aid to students."

Of Volume III., Part I.

THE SCOTSMAN says:—"It is a valuable little book and deserves the attention both of teachers and of students of the subject."

In 2 vols. Globe 8vo.
Practical Physics for Schools and Junior Students of Colleges.
By BALFOUR STEWART, M.A, LL.D., F.R.S., late Langworthy Professor of Physics, Owens College, Manchester, and Professor of the Victoria University, and W. H. HALDANE GEE, Demonstrator and Assistant Lecturer in Physics, the Owens College.

Vol. I. Electricity and Magnetism. 2s. 6d.
Vol. II. Optics, Heat and Sound. *[In preparation.*

Of Volume I.

THE BRITISH MEDICAL JOURNAL says:—"This little work will be welcomed by teachers of physics in schools. Each chapter is subdivided into lessons of suitable length for school instruction, whilst the lessons contain exercises, and laws are illustrated by examples. All apparatus required in experiments is fully described, and even the mode of its manufacture is carefully detailed, so that the student may first make his apparatus and then apply it to experimental purposes. In the opinion of the authors, the greater part of the work should be easily within the range of schoolboys. The eminently practical nature of the treatise is maintained to the end."

THE ATHENÆUM says:—"This is a very serviceable guide to laboratory work in the subjects indicated, written from the point of view of the workshop as much as of the lecture-room. It will doubtless attract the attention of all competent teachers of Practical Physics."

Foolscap 8vo. 3s. 6d.
Elementary Lessons on Sound.
By Dr. W. H. STONE, Lecturer on Physics at St. Thomas's Hospital. With Illustrations.

NATURE.—"Gives in a very small compass a large amount of information as to the laws and phenomena of sound. The author has not only extracted the essence of what is contained in bulky and expensive treatises, sometimes in foreign languages, but he has also given much additional information from memoirs and transactions of scientific societies out of the reach of the ordinary public. The application of acoustics to musical instruments is a useful addition, the subject being one which the author has made specially his own. He has also stated some of the simplest facts of the connection between acoustical phenomena and the structure of music."

THE ENGLISH MECHANIC.—"This is an excellent little text-book on sound, well adapted for students of the Science and Art classes, but also of considerable value to those who desire to do something more than cram."

Foolscap 8vo. 4s. 6d.
Elementary Lessons in Electricity and Magnetism.
By SILVANUS P. THOMPSON, D.Sc., F.R.S., Principal and Professor of Physics in the Technical College Finsbury. With Illustrations. New Edition.

THE LANCET.—"A Standard Text-Book on the subject."

THE EDUCATIONAL TIMES.—"The book is really, as it claims to be, up to date, and is indeed the only text-book for genuine students a grade above the elementary."

THE SATURDAY REVIEW.—"This is a new edition, practically the eighteenth, of this most valuable text-book. It has been brought up to date both in practice and in theory, and we see every reason to expect that this issue will be exhausted as rapidly as have been previous issues."

THE ELECTRICIAN.—"The favourite text-book among readers approaching the subject from the pure science side, and wishing to obtain a good grounding in both the pure and applied parts of the subject."

8vo. 6s. 6d. net. Mathematical Tables, separately, 1s. 6d.

Mathematical and Physical Tables. For the use of Students in Technical Schools and Colleges. By JAMES P. WRAFSON, B.A. (Dublin), Chief Lecturer in Pure and Applied Mathematics at the Municipal Technical School, Manchester, and W. W. HALDANE GEE, B.Sc. (Lond.), Chief Lecturer in Physics and Electrical Engineering at the Municipal Technical School, Manchester; formerly Lecturer of the Victoria University.

PHYSICS.
ADVANCED.

8vo. 21s.

Physics: Advanced Course. By GEO. F. BARKER, Professor of Physics in the University of Pennsylvania. With a magnetic map of United States.

THE NATION.—"The arrangement of subjects treated, and the names given them make the work specially noteworthy as an exponent of modern views in physics. While these views are not in themselves new, they here form what we believe is essentially a new classification of the subject for purposes of study and instruction based on the most recent views of energy considered as being ultimately a phenomenon of the author."

THE CHEMICAL NEWS.—"We can scarcely hesitate to pronounce this work the best truly modern manual of physics in our language."

Medium 8vo. 21s.

A Text-Book of the Principles of Physics. By ALFRED DANIELL, M.A., LL.B., D.Sc., F.R.S.E., of the Inner Temple, Barrister-at-Law, Member of the Faculty of Advocates in Scotland, formerly Lecturer on Physics in the School of Medicine, Edinburgh. Third Edition (Seventh Thousand).

THE LANCET.—"For the first M.B. Lond., students should possess a copy of . . . A Text-book of Physics."

THE EDINBURGH MEDICAL JOURNAL.—"This volume supplies a distinct want in scientific literature, and from the point of view of profession it is so interesting that it merits a place in the library of every medical practitioner and student."

Globe 8vo. 5s.

Illustrations of the C.G.S. System of Units. With Tables of Physical Constants. By J. D. EVERETT, M.A., D.C.L., F.R.S., F.R.S.E., Professor of Natural Philosophy, Queen's College, Belfast. Fourth Edition.

THE ACADEMY.—"It must form an essential component of the libraries of all physical laboratories, and the investigator in physical science will find it one of his most useful companions. The arrangement is good, the work is conscientiously done, and a great mass of useful information has been collected from very diverse sources."

Foolscap 8vo. 5s. 6d.

Absolute Measurements in Electricity and Magnetism.
By ANDREW GRAY, M.A., F.R.S.E., Professor of Physics in the University College of N. Wales. Second Edition.

THE ACADEMY.—"We cordially recommend it to all students of electricity and magnetism who are not sufficiently advanced to fully profit by the author's larger work on the same subject."

INDUSTRIES.—"The student will find in it a clear definition of the various absolute units used, minute accounts of the methods of determining them, and careful descriptions of the instruments used in measuring electrical and magnetic quantities in terms of them. Besides this the book abounds in practical hints respecting the *minutiæ* of laboratory work which make all the difference between accurate and worthless experiment."

NATURE.—"Is a most interesting book to read."

Medium 8vo.

Magnetism and Electricity.
By ANDREW GRAY, M.A., F.R.S.E., Professor of Physics in University College of N. Wales, Bangor. Vol. I. 14s. net.

[Vol. II. *in the Press.*

Crown 8vo. 6s. 6d.

Modern Views of Electricity.
By OLIVER J. LODGE, D.Sc., LL.D., F.R.S., Professor of Experimental Physics in University College, Liverpool. With Illustrations.

THE ELECTRICIAN.—"Gathering up from the researches of the most able and modern of investigators down to within two months of the date of publication and adding to them the result of his own admirable and ingenious inquiries, he has now strung together and expounded in language as graphic as it is simple, those modern views of electricity which will assuredly in the main survive the present orthodox notions."

THE ELECTRICAL ENGINEER.—"Every practical man as well as every student should examine it. To our mind there is no living exponent of these theoretical views so able as the author of this book."

THE EDUCATIONAL TIMES.—"The book is altogether one of the most original and most suggestive books for thinking workers in this subject, which the last few years have produced."

8vo. 8s. 6d. net.

A Laboratory Course in Experimental Physics.
By W. G. LOUDON, B.A., Demonstrator in Physics in the University of Toronto, and G. C. McLENNAN, B.A., Assistant Demonstrator in Physics in the University of Toronto.

NATURE.—"The book is divided into two parts. Part I. is devoted to the description of elementary apparatus and elementary methods of performing simple experiments in mechanics, light, photometry and heat. . . . . Part II. is intended for advanced students and contains directions for experiments in acoustics, heat, electricity, and magnetism, with an appendix on the determination of gravity and on the torsion pendulum. . . . The choice of experiments is good, the theory and descriptions of the apparatus are accurate, and the illustrations are clear. . . . . The book is well got up, and its value is greatly enhanced by the tables of physical constants at the end. We heartily recommend the work to teachers of practical physics as containing very good matter, and being accurate and free from misprints."

THE EDUCATIONAL REVIEW.—" The general style of the book is good. It is copiously illustrated, and the well-worn fundamental experiments are treated freshly and clearly."

8vo. 12s. 6d. net.

A Laboratory Manual of Physics and Applied
Electricity. Arranged and Edited by EDWARD L. NICHOLS, Professor of Physics in Cornell University. In Two Volumes. Illustrated.

(For Vol. I. see under Physics, Elementary.)

Vol. II. Senior Courses and Outlines of Advanced Work. By GEORGE S. MOLER, FREDERICK BEDELL, HOMER G. HOTCHKISS, CHARLES P. MATTHEWS, and the EDITOR.

THE OXFORD MAGAZINE.—" We know no text-book in English which covers the same ground, and can cordially recommend it to the notice of teachers and students."

THE CAMBRIDGE REVIEW.—" The chapter on alternating current experiments deals with matter that seldom appears in physical text-books, and will be found most useful."

THE INDIAN ENGINEER.—" Some years ago we welcomed the able work of Professor Balfour Stewart and Mr. Haldane Gee (see page ..). . . . . Now we have an equally good guide for second and third year men, and what we never before possessed, a reference book for those about to proceed to original work."

Crown 8vo. 5s.

Intermediate Course of Practical Physics. By
ARTHUR SCHUSTER, F.R.S., Professor of Physics in the Owens College, Manchester, and CHARLES H. LEES, D.Sc.

THE JOURNAL OF EDUCATION.—" The experiments given are well detailed. . . . the directions are concise, and pitfalls are pointed out."

THE GUARDIAN.—" The book is therefore valuable not only for the 'Intermediate B.Sc. and first M.B. examinations' of the modern Universities, but is also useful at the older Universities, especially under the new plans where some practical work in physics is required as a preliminary. The work covers mechanics and physics, including heat, light, sound, magnetism, electric currents and electric charges in the order of parts named. . . . . The volume will undoubtedly be of great use beyond the precincts of the Owens College. It will lighten the burden of teachers and demonstrators in writing or lithographing instructions; it will also give them the benefit of the very extensive experience of the able staff at the Owens College."

BIOLOGY.

ELEMENTARY.

Pott 8vo. 1s.

First Lessons in Practical Botany. By G. T. BETTANY, M.A., B.Sc., F.L.S., Lecturer on Botany in Guy's Hospital Medical School.

NATURE.—"This is an excellent little book. Its diligent study by teachers as well as pupils would give descriptive botany the real educational value which is so often claimed for it and which, at the bottom, it no doubt possesses, if only the old type of manuals could be exterminated."

Globe 8vo. 3s. 6d.

Practical Botany for Beginners. By F. O. BOWER, D.Sc., F.L.S., Regius Professor of Botany in the University of Glasgow.

THE JOURNAL OF EDUCATION.—"Many improvements have been made in the book since its first appearance, and it is only justice to say that as it was their the only book of the kind in the language, so it is now the best and most complete."

THE GLASGOW HERALD.—"Some valuable extensions and rearrangements have been made. The student can now study the work in a consecutive manner, and the full directions given in the earlier chapters will aid in furnishing the laboratory in a thoroughly practical manner."

Pott 8vo. 1s.

Botany. By Sir J. D. HOOKER, K.C.S.I., C.B., M.D., D.C.L. Oxon., LL.D. Cantab., F.R.S., F.L.S., F.G.S.

[*Science Primers.*

THE FIELD.—"The most compact introduction to botany which we have seen. ... We have seldom met with a more suggestive and interesting little volume, or one which is more calculated to lead the student on to more advanced works, providing as it does, a solid basis upon which a more extensive superstructure may at any time be raised."

THE AGRICULTURAL GAZETTE.—"There ought to be twenty thousand purchasers of such a book as this to begin with—in the agricultural ranks alone."

Crown 8vo. 6s.

The Living Organism. An Introduction to the Problems of Biology. By ALFRED EARL, M.A., late Scholar of Christ's College, Cambridge, of the Inner Temple, Barrister-at-Law.

opportune volume. That it will have an influence upon the biological work of the future we have not the slightest doubt, and we believe also that its influence will be great, permanent and beneficial."

NATURE.—"The whole work must be carefully read by every serious student; and there can be no question of its great and permanent value, as a contribution to our knowledge of a particular class of variations and as a stimulus to further work."

Crown 8vo. Price 10s. 6d.

A Course of Practical Instruction in Botany.
By F. O. BOWER, D.Sc., F.L.S., Regius Professor of Botany in the University of Glasgow.

THE SATURDAY REVIEW.—"There can be little doubt that this book has now established its claims to be regarded as *the* course of practical botany in this country."

NATURE.—"The extensive changes which this volume has undergone have rendered it more than ever an invaluable aid to the study of plants in the laboratory. English students may be congratulated on their good fortune in possessing such a hand-book, and we may confidently hope that the present edition of Professor Bower's work may render even greater services to scientific education than did its predecessor."

Medium 8vo. 17s. net each volume.

THE CAMBRIDGE NATURAL HISTORY.
Edited by S. F. HARMER, M.A., Fellow of King's College, Cambridge, and Superintendent of the University Museum of Zoology; and A. E. SHIPLEY, M.A., Fellow of Christ's College, Cambridge, University Lecturer on the Morphology of Invertebrates. Illustrated.

Vol. II. FLATWORMS AND MESOZOA. By F. W. GAMBLE, M.Sc. (Vict.), Owens College. NEMERTINES. By Miss L. SHELDON, Newnham College, Cambridge. THREAD-WORMS AND SAGITTA. By A. E. SHIPLEY, M.A., Fellow of Christ's College, Cambridge. ROTIFERS. By MARCUS HARTOG, M.A., Trinity College, Cambridge, (D.Sc. Lond.), Professor of Natural History in Queen's College, Cork. POLYCHAET WORMS. By W. BLAXLAND BENHAM, D.Sc. (Lond.), Hon. M.A. (Oxon.), Aldrichian Demonstrator of Comparative Anatomy in the University of Oxford. EARTHWORMS AND LEECHES. By F. E. BEDDARD, M.A. (Oxon.), F.R.S., Prosector to the Zoological Society, London. GEPHYREA AND PHORONIS. By A. E. SHIPLEY, M.A., Fellow of Christ's College, Cambridge. POLYZOA. By S. F. HARMER, M.A., Fellow of King's College, Cambridge.

Vol. III. MOLLUSCS AND BRACHIOPODS. By Rev. A. H. COOKE, M.A., Fellow of King's College, Cambridge. BRACHIOPODS (Recent). By A. E. SHIPLEY, M.A., Fellow of Christ's College, Cambridge. BRACHIOPODS (Fossil). By F. R. C. REED, B.A., Trinity College, Cambridge.

Vol. V. PERIPATUS. By ADAM SEDGWICK, M.A., F.R.S., Fellow and Lecturer of Trinity College, Cambridge. MYRIAPODS. By F. G. SINCLAIR, M.A., Trinity College, Cambridge. INSECTS. Part I. By DAVID SHARP, M.A. Cantab., M.B. Edin., F.R.S.

Vol. IX. BIRDS. By A. H. EVANS, M.A.

Of Volume II.

THE CAMBRIDGE REVIEW says:—" Most of the articles are of a very high order of merit. Taken as a whole it may be said that they are by far the best which have as yet been published.... We may say with confidence that the same amount of information within the same compass is to be had in no other zoological work."

Of Volume III.

THE BRITISH MEDICAL JOURNAL says:—" The great expectations we have been led to form have in this volume ... been realised.... It is the product of the united labours of the leading men of a great zoological school, and as such must take a high place."

THE EDINBURGH REVIEW says:—" Much of it we have read with pleasure, and much of it with profit, deriving satisfaction in the one case from the process, in the other from the accomplished fact."

Of Volume V.

THE ENTOMOLOGICAL MONTHLY MAGAZINE says:—" We venture to think the work will be found indispensable to all who seek to extend their general knowledge beyond the narrowing influence of exclusive attention to certain Orders or Groups, and that it will take a high position in 'The Cambridge Natural History' series."

THE FIELD says:—" To students who wish to know more than the mere superficial aspects of the vast extent of zoology, the work is one of an invaluable character.... Although written for the student and the specialist the book is not less adapted to all intelligent readers who wish to make themselves thoroughly acquainted with the habits, structure, and the modern classification of the animals of which it treats. To such it cannot be recommended too strongly."

New Edition. 8vo. 10s. 6d.

The Botanical Text-Book. Part I. Structural Botany: or Organography on the Basis of Morhpology. To which is added the Principles of Taxonomy and Phytography, and a Glossary of Botanical Terms. By ASA GRAY, LL.D., Fisher Professor of Natural History (Botany) in Harvard University. Sixth Edition.

THE BRITISH MEDICAL JOURNAL.—" Is widely known as one of the best [Botanical Text-books] in the English language."
NATURE.—" A valued hand-book."

Globe 8vo. 10s. 6d.

The Student's Flora of the British Islands. By Sir J. D. HOOKER, K.C.S.I., C.B., M.D., D.C.L. Oxon., LL.D. Cantab., F.R.S., F.L.S., F.G.S. Third Edition.

THE GARDEN.—" Too much praise cannot be bestowed on the general design of the book.... We may enumerate a few of the particularly good points of the work in hand. The essential characters and properties of the Natural Orders and genera are given, their distribution, the number of genera in each Natural Order and the natural species in each genus."

In 2 Parts. 8vo. 17s. net. per Part.

A Text-Book of Comparative Anatomy. Invertebrates only. By Dr. ARNOLD LANG, Professor of Zoology in the University of Zurich. With Preface to the English Translation by Professor Dr. ERNST HÆCKEL, Director of the Zoological Institute in Jena. Translated into English by HENRY M. BERNARD, M.A., and MATILDA BERNARD.

NATURE.—"In a short notice it is quite impossible to give any idea of the interesting way in which the book is written. The English translation is all that could be desired; the illustrations are excellent. The arrangement of the subject matter has been carefully thought out and reference to any subject is assisted by the use of different kinds of type in the text. A long classified list of the important literature is given at the end of each chapter."

THE TIMES.—"The book is a monument of labour: and the reputation of the author is a sufficient guarantee for the general accuracy of his statements."

Large 8vo. 25s.

Human Embryology. By CHARLES SEDGWICK MINOT, Professor of Histology and Human Embryology, Harvard Medical School, Boston. With 463 Illustrations.

THE LANCET.—"This well got up and richly illustrated work is the result, the author tells us, of ten years' labour, and is without doubt the fullest and most complete account of Embryology of man that has hitherto appeared in the English language. . . . The treatise may be strongly recommended as an excellent exposition of the present state of embryological knowledge."

NATURE.—"The work before us constitutes a monument of erudition in the difficult and complex subject of which it treats."

Crown 8vo. 8s. 6d.

A Course of Instruction in Zootomy. Vertebrata. By T. JEFFERY PARKER, B.Sc., F.R.S., Professor of Biology in the University of Otago, New Zealand. With Illustrations.

THE SATURDAY REVIEW.—"An admirable work. . . . In large and well appointed morphological laboratories, such as Professor Huxley's own, or that which the late Professor Balfour established at Cambridge, this sort of help is supplied by a staff of demonstrators; but Mr. Parker has conferred an immense boon on teachers and students alike, by enabling the latter to dispense to a very great extent with the services of the former. He has selected half a dozen types of vertebrate structure—the lamprey, the skate, the cod, the lizard, the pigeon and the rabbit, and under each head gives clear directions as to the way in which these creatures are to be dissected, and what points the students ought to make out at the successive stages of his work."

8vo. 8s. net.

The Structure of Man: an Index to his Past History. By ROBERT WIEDERSHEIM, Professor of Anatomy, and Director of the Institute of Human and Comparative Anatomy in the University of Freiburg, in Baden. Translated

by H. and M. BERNARD. The Translation, edited and annotated and a preface written by G. B. HOWES, F.L.S., Professor of Zoology, Royal College of Science, London. With 105 figures in the text.

THE EDINBURGH MEDICAL JOURNAL.—"Viewed as a series of problems in evolution, every part [of the study of human anatomy] is of entrancing interest. It is in this aspect that Wiedersheim presents ' The Structure of Man.' We are briefly taken over the whole bodily structure, and the evidence of progressive and retrogressive changes is lucidly pointed out. The student who cannot derive pleasure from such a study must be dull and unimaginative indeed. It is needless here to analyse the work in detail. Every part is worthy of careful study. The work of the translators has been well done, and Professor Howes's annotations are of great value."

THE CAMBRIDGE REVIEW.—"The book is one which should be in the hands of every Anthropologist and read by every medical student."

Medium 8vo. 12s. 6d. net.

Elements of Comparative Anatomy of Vertebrates.
Adapted from the German of ROBERT WIEDERSHEIM, Professor of Anatomy and Director of the Institute of Human and Comparative Anatomy in the University of Freiburg, in Baden. By W. NEWTON PARKER, Professor of Biology in the University College of S. Wales and Monmouthshire. With Additions by the Author and Translator. 2nd Edition. 333 Woodcuts.

EXTRACT FROM PREFACE TO THE FIRST EDITION.—"The arrangement of the book according to organs, and not according to groups of animals, is likely to render it more difficult for a beginner, and a general knowledge of Zoology will be of great assistance. The pages on which the different groups are described are, however, collected together in the index, so that the sections relating to any one group can be easily referred to. The present arrangement seems to be the only possible one if the book is to be founded on a scientific basis, for it is most important that the student should grasp the fact that there has been an evolution of organs, as well as of animals.

*   *   *   *   *

"The book was written for students of Medicine, as well as for those of Comparative Anatomy : the intimate connection of the two subjects renders it most important that medical students should have a general scientific basis for their special anatomical knowledge."

EXTRACT FROM PREFACE TO THE SECOND EDITION.—"The book has been almost entirely rewritten, and I had the advantage of revising the greater part of the manuscript cast, and of submitting the proof-sheets to Professor Wiedersheim, who has furnished me with much new material. A number of the old figures have been replaced and several additional ones inserted."

THE SATURDAY REVIEW.—" Professor Wiedersheim and his translator are to be congratulated on this work, for which they are jointly responsible. The descriptions of the structure and development of the organ-systems of vertebrates, though necessarily brief, are clear, and their meaning is made more apparent by a large number of excellent woodcuts."

THE JOURNAL OF EDUCATION.—" Messrs. Macmillan and Co. are to be congratulated upon not merely keeping the book up to the standard of such of their text-books as Professors Balfour and Ziegler's works, but actually improving on them."

Vol. II. Medium 8vo. 18s.

A Text-Book of the Physiological Chemistry of
the Animal Body, including an account of the chemical changes occurring in disease. By ARTHUR GAMGEE, M.D., F.R.S., Brackenbury Professor of Physiology in the Owens College. Vol. II. The Physiological Chemistry of Digestion. With Illustrations.

THE LANCET.—" The best manual on the subject . . . for advanced students."

THE EDINBURGH MEDICAL JOURNAL.—" Undoubtedly the best and most exhaustive treatise on the subject."

THE BRITISH MEDICAL JOURNAL.—" Will prove of inestimable value to the student, to the practitioner and to the investigator."

Pott 8vo. 4s. 6d.

Lessons in Elementary Physiology. By Professor
T. H. HUXLEY, F.R.S. Illustrated. Fourth Edition.

THE MEDICAL PRESS.—" It is an admirable illustration of how the greatest masters of a science may teach its elements in the most simple manner."

THE WESTMINSTER REVIEW.—" Unquestionably the clearest and most complete elementary treatise on this subject that we possess in any language. In this admirable little work the outlines of the physiology of the human body are set forth in the plainest English, without any of the endeavours of fine writing in which a French author would certainly have indulged, but with a simplicity and earnestness of purpose which command our highest admiration."

Pott 8vo. 1s. 6d.

Questions on Huxley's Lessons in Elementary
Physiology for the use of Schools. By THOMAS ALCOCK, M.D. Fifth Edition.

THE GUARDIAN.—" Will be found a useful auxiliary to the teacher."

THE MUSEUM.—" This is a series of very minute questions on Huxley's admirable little work on Physiology. It will be useful for those teachers who are in the unfortunate position of not being able to set questions themselves."

8vo. 10s. net.

A Text-Book of Anatomy and Physiology for
Nurses. Compiled by DIANA CLIFFORD KIMBER, Graduate of Bellevue Training School, Assistant Superintendent N. Y. City Training School, Blackwell's Island, N. Y.

THE LANCET.—" This work is written in a clear and easy style, and is fully illustrated with excellent engravings."

GUY'S HOSPITAL GAZETTE.—" As a book of reference we have nothing but praise for the volume before us. The diagrams are excellent and well chosen, whilst the matter is clearly expressed."

THE BOSTON LITERARY WORLD.—" Is admirably adapted for its purpose of giving facts most important for the nurse, and it is as well adapted for use by young women in school or college."

Medium 8vo.

General Physiology. An outline of the Science of Life. By MAX VERWORN, M.A., Ph.D., A.O., Professor of Physiology in the Medical Faculty of the University of Jena. Translated from the Second German Edition, and edited by FREDERIC S. LEE, Ph.D., Adjunct Professor of Physiology in Columbia University. With 285 illustrations.

PATHOLOGY AND BACTERIOLOGY.

In 2 vols. Medium 8vo.

A Text-Book of Pathology. Systematic and Practical. By D. J. HAMILTON, M.B., F.R.C.S.E., F.R.S.E., Professor of Pathological Anatomy (Sir Erasmus Wilson Chair), University College of Aberdeen. Copiously illustrated. Vol. I. 21s. net.

Vol. II. Parts I. and II. 15s. net, each part.

THE LANCET.—"The most complete and systematic work published in the English language. Contains a very full and accurate bibliography. Invaluable for reference and for study for the higher examinations."

THE EDINBURGH MEDICAL JOURNAL.—"The book is a great work. That it should take its place as the first text-book of pathology there can be little doubt. . . . . To the advanced student and worker in pathology it should prove of inestimable value."

8vo. 6s.

Methods of Pathological Histology. By C. VON KAHLDEN, Assistant Professor in University of Freiburg. Translated by H. MORLEY FLETCHER, M.A., M.D. With an Introduction by G. SIMS WOODHEAD, M.D.

THE BRITISH MEDICAL JOURNAL.—"This book will prove most valuable to all pathologists. It puts in a short space and in a clear and lucid form the principal histological methods at present in use, including those which have been recently introduced. Dr. Woodhead says in the introduction that Dr. Morley Fletcher has conferred a real boon on the busy student of morbid histology by translating this well known and highly valued work of Professor v. Kahlden. With this we entirely agree. The translation is well done and the editor's footnotes, many of which are inspired by Professor Sherrington, give additional hints and information which will render the book more valuable than a bare translation would have been."

Crown 8vo. 4s. 6d.

A Course of Elementary Practical Bacteriology. Including Bacteriological Analysis and Chemistry. By A. A. KANTHACK, M.D., M.R.C.P., and J. H. DRYSDALE, M.B., M.R.C.P.

THE LANCET.—"A guide to the elementary study of bacteriology. The work deals in turn with general and special methods of cultivation and examination of micro-organisms, with the more important species of bacteria, and then with bacteriological analysis and bacteriological chemistry."

Crown 8vo. 5s.

The Bacteria in Asiatic Cholera. By E. KLEIN, M.D., F.R.S., Lecturer on General Anatomy and Physiology in the Medical School of St. Bartholomew's Hospital, Professor of Bacteriology at the College of State Medicine, London.

THE LANCET.—" No single observer has contributed more to the knowledge on this subject than Dr. Klein, and the work before us is a thoroughly scientific exposition of it in all its bearings. . . . . A perusal of his book cannot fail to leave the impression that there is still much to be learnt concerning cholera and its relationship to bacterial life."

NATURE.—"The present volume is a valuable and most trenchant criticism of every step of Koch's argument, and may be said to contain everything that can at present be said against Koch's theory, of which the author is the most active opponent."

Crown 8vo. 10s. 6d.

Mirco-Organisms and Disease : An Introduction into the study of specific micro-organisms. By E. KLEIN, M.D., F.R.S., Lecturer on General Anatomy and Physiology in the Medical School of St. Bartholomew's Hospital, Professor of Bacteriology at the College of State Medicine, London. New Edition. Revised. With 201 Illustrations.

THE LANCET.—"A standard text-book."

THE BRITISH MEDICAL JOURNAL.—"Ten years have elapsed since the last edition of this book appeared, and the new edition is to all intents and purposes a new work, so large have been the advances in bacteriology during that time. . . . . We have conscientiously read the book from preface to index and can unhesitatingly speak of it in terms of respectful admiration. . . . . The book, from the matter which it contains, should find a place in the library of every person interested in the question of bacteriology, from the botanist to the medical officer of health."

In 3 Parts. Medium 8vo.

A Text-Book of Pathological Anatomy and Patho-Genesis. By ERNST ZIEGLER, Professor of Pathological Anatomy in the University of Tübingen. Translated and edited for English Students by DONALD MACALISTER, M.A., M.D., Fellow of the Royal College of Physicians, Fellow and Medical Lecturer of St. John's College, University Lecturer in Medicine.

Part I. General Pathological Anatomy. Second Edition. 12s. 6d.

Part II. Special Pathological Anatomy. Sections I.-VIII. 17s. net.

Part III. Special Pathological Anatomy. Sections IX.-XV. 17s. net.

THE LANCET. — "A good text-book for the higher examinations. . . . . References to literature very full."

THE BRITISH MEDICAL JOURNAL.—"It is certainly the most complete account of this branch of medical science which exists. The work is too well known to require further notice at our hands, and we can only congratulate the editors on having placed it within the reach of English readers."

PHARMACOLOGY AND THERAPEUTICS.

8vo. In One Vol. 21s. In Two Vols. 22s. 6d.

A Text-Book of Pharmacology, Therapeutics and Materia Medica. By T. LAUDER BRUNTON, M.D., D.Sc., F.R.S., Fellow of the Royal College of Physicians; Assistant Physician and Lecturer on Materia Medica at St. Bartholomew's Hospital; Examiner in Materia Medica in the University of Oxford; late Examiner in the Universities of Edinburgh and London, in the Victoria University, and in the Royal College of Physicians, London. Adapted to the United States' Pharmacopœia by F. H. WILLIAMS, M.D. Third Edition, containing the additions, 1891, to the British Pharmacopœia.

>   *THE LANCET.*—"The student for the higher examinations would be well advised to provide himself with this work. It is equally useful as a practical treatise and as a work of reference, and is admirably written."
>
>   *THE BRITISH MEDICAL JOURNAL.*—"We feel that we cannot speak too highly of Dr. Brunton's work. It is worthy of his high reputation as a pharmacologist, and is undoubtedly the best treatise on the subject in the English language."

Crown 8vo. 6s.

Pharmacology and Therapeutics: or Medicine, Past and Present. The Goulstonian Lectures delivered before the Royal College of Physicians in 1877. By T. LAUDER BRUNTON, M.D., D.Sc., F.R.S.

>   *NATURE.*—"The book is one which may be studied with benefit by all medical men, and those not belonging to the profession who desire to have an insight into some of the tendencies of modern medicine will find it a trustworthy and intelligent guide."
>
>   *THE WESTMINSTER REVIEW.*—"The volume will be read with equal interest both by members of the medical profession and the lay public, as it contains a large amount of information which is conveyed in an easy and pleasant style."

8vo. 10s. 6d. net.

The Action of Medicine. Being the Course of Lectures on Pharmacology and Therapeutics delivered at St. Bartholomew's Hospital during the Summer Session of 1896. By T. LAUDER BRUNTON, M.D., D.Sc., F.R.S.

>   *THE TIMES.*—"Lectures on Pathology and Therapeutics are the course delivered by the author in the Medical School of St. Bartholomew's Hospital, and deal very completely with the various actions exerted by medicines upon the human body. They necessarily appeal chiefly to medical readers, but are nevertheless calculated to dispel many errors which cling to the minds of the general public. Dr. Lauder Brunton possesses in an eminent degree the power of adorning his explanations of principles by interesting and graphic illustrations of their applicability to the events of life; and at least among the junior members of the medical profession, there will be few who might not learn from his pages many things of great practical importance."

8vo. 3s. 6d. net.

An Introduction to Modern Therapeutics.
By T. LAUDER BRUNTON, M.D., D.Sc., F.R.S.; Delivered before the Royal College of Physicians in London. June, 1889.

THE MEDICAL PRESS.—" The matter composing these lectures, as we would expect from their author, is judiciously selected, and fully brought up to the highest standard of scientific research. . . . The ways in which the principal classes of remedies exercise their respective influences on the system are in every instance thoroughly examined; and we feel sure that, for many readers, facts and phenomena which had always been looked upon as insoluble mysteries—although witnessed every day—are now placed in an open and familiar light. We need hardly add that we cordially recommend this volume to every earnest student of medicine."

Pott 8vo. 3s. 6d.

Lessons on Prescriptions and the Art of Prescribing.
By W. HANDSEL GRIFFITHS, PH.D., L.R.C.P.E., Licentiate of the Royal College of Surgeons, Edinburgh; Professor of Chemistry in the Ledwich School of Medicine, Dublin; etc.

GUY'S HOSPITAL GAZETTE.—" We can commend the work to the attention of the student; and candidates for the Hall will find it specially useful, and we believe the practitioner will find in it many a useful reference in cases of difficulty."

THE PHARMACEUTICAL JOURNAL.—" We think this little book better suited for educational as well as examination purposes than any other of the same size with which we are acquainted."

Crown 8vo. 8s. 6d.

A Text-Book of General Therapeutics.
By W. HALE WHITE, M.D., F.R.C.P., Senior Assistant Physician to, and Lecturer on Materia Medica and Therapeutics at, Guy's Hospital. With Illustrations.

THE BRITISH MEDICAL JOURNAL.—" There are very readable and well written articles on climatology, massage, baths, electricity, hypnotism and other similar means of treating disease. . . . Dr. White must be considered to have done a great service to junior practitioners in putting together a short and trustworthy account of what is known regarding these subjects."

THE HOSPITAL.—" Contains a wealth of information, including all he most recent developments. From medical climatology to hypnotism is a long journey, but the reader will find his path made easy and pleasant, and cannot fail to profit much, though he travels fast. This book is welcome and timely. It has been written by a master hand, and will no doubt have a wide popularity. . . . We know of no text-book yet published in which the outlines of modern views and modern methods are so fully and so concisely represented."

GYNÆCOLOGY.

Medium 8vo. 25s. net.

A System of Gynæcology. By MANY WRITERS. Edited by T. CLIFFORD ALLBUTT, M.A., M.D., LL.D., F.R.C.P., F.R.S., F.L.S., F.S.A., Regius Professor of Physic in the University of Cambridge, etc., etc., and W. S. PLAYFAIR, M.D., F.R.C.P.

THE QUARTERLY MEDICAL JOURNAL.—"The volume in its entirety is well worthy the perusal of all who have a rational interest in gynæcological practice."

THE GLASGOW MEDICAL JOURNAL.—"The names by which each essay individually is signed are sufficient guarantee of adequate treatment.... This work may be safely used as a standard of reference, for learning details of modern surgical operations."

SURGERY.

Royal 8vo. 10s. net.

A Treatise on the Ligation of the Great Arteries in Continuity. With Observations on the Nature, Progress and Treatment of Aneurism. By CHARLES A. BALLANCE, M.B., M.S. Lond., F.R.C.S., Assistant Surgeon to St. Thomas's Hospital, and WALTER EDMUNDS, M.A., M.C. Cantab., F.R.C.S., Resident Medical Officer, St. Thomas's Home. Illustrated by 10 plates and 232 figures.

THE BRITISH MEDICAL JOURNAL.—"The manner in which their researches have been carried on, and the light they have thrown on many points of purely scientific interest, cannot fail to establish the reputation of the authors as able and earnest investigators."

THE LANCET.—"No labour has been spared to make the investigation thorough and complete.... The result is a volume highly creditable to its authors.... The candid reader will be slow to question the results arrived at by such evidently painstaking and impartial observers."

Crown 8vo. 10s. 6d. net.

A Treatise on Diseases of the Nose and its Accessory Cavities. By GREVILLE MACDONALD, M.D. (Lond.), Physician to the Hospital for Diseases of the Throat. Second Edition.

8vo. 15s. net.

A Clinical Treatise on Diseases of the Breast. By A. MARMADUKE SHEILD, M.B. (Cantab.), F.R.C.S.,

8vo. 17s. net.

A Manual of Public Health. By A. WYNTER BLYTH, M.R.C.S., L.S.A., Fell. Chem. Soc., Fell. Inst. Chem., Barrister-at-Law, Medical Officer of Health and Public Analyst for St. Marylebone, etc.

THE LANCET.—"An excellent manual dealing with general principles. It is fairly exhaustive and affords a good book of reference."

NATURE.—"Mr. Wynter Blyth may be congratulated on the excellent text-book he has produced, based as it is upon the practical experience of many years of sanitary work, obtained in one of the largest metropolitan districts . . . of the high value of the work as a text-book of public health there can be no question; and we hope that Mr. Blyth's manual will be in the hands not only of students, but of all those whose calling is sanitary science."

8vo. 8s. 6d. net.

Lectures on Sanitary Law. By A. WYNTER BLYTH, M.R.C.S., L.S.A., Fell. Chem. Soc., Fell. Inst. Chem., Barrister-at-Law, Medical Officer of Health and Public Analyst for St. Marylebone, etc.

NATURE.—"This work presents a general view of the powers and duties of Local Authorities to public health, and since the material has been compiled by one who, while he is a prominent sanitarian is also a barrister-at-law, the fact that the work is good and trustworthy, and leaves but little to be desired, goes without saying."

THE TIMES.—"A lucid and capable exposition of the outlines of a great and growing subject . . . . . setting forth in small compass a general view of the powers and duties of Local Authorities in relation to public health."

8vo. 16s.

The Treatment and Utilisation of Sewage. By W. H. CORFIELD, M.A., M.D. Oxon., Fellow of the Royal College of Physicians, London; Professor of Hygiene and Public Health at University College, London; Ex-President of the Society of Medical Officers of Health. Third Edition, revised and enlarged by the Author and LOUIS C. PARKES, M.D., Cert. Public Health, London.

THE BRITISH MEDICAL JOURNAL.—"Dr. Corfield's work deserves attentive study by all on whom the responsibility of sewerage and sewage disposal may devolve, and will be found indispensable to the working sanitarian. With such a guide to what has been attempted and what may be achieved, it will be a disgrace to our metropolitan and urban authorities, if they allow German and Italian municipal boards to prove themselves more enlightened than they are."

THE MEDICAL PRESS—"The importance and scope of the subject and the authority of the author render further commendation unnecessary. The work is an indispensable addition to the library of sanitary engineers, medical officers of health, and others interested in these matters."

Pott 8vo. 1s.

On Preservation of Health in India. By Sir JOSEPH FAYRER, K.C.S.I., M.D., F.R.S., President of the Medical Board at the India Office.

THE BRITISH MEDICAL JOURNAL.—"It is a concise and useful description of the conditions—geographical, climatic, meteorological and otherwise—which are encountered in the various parts of the Indian Empire. . . . . Persons proceeding to India or other tropical countries would do well to 'read, mark, learn and inwardly digest' the excellent precepts contained in this admirable little book."

THE LANCET.—"Sir Joseph Fayrer's long medical experience and thorough knowledge of India guarantee the soundness of his recommendations."

Crown 8vo. 3s. 6d.

The Soil in Relation to Health. By HENRY A. MIERS, F.R.S., Professor of Mineralogy in the University of Oxford, and ROGER CROSSKEY, M.A., D.P.H., Fellow of the British Institute of Public Health. With Illustrations.

THE BRITISH MEDICAL JOURNAL.—"A pleasantly written, well-arranged, and instructive little work, which may be studied with profit by those preparing for public health examinations, and others."

THE LANCET.—"It would greatly conduce to the progress of sober and rational sanitary reform if the information contained in this volume became part of an ordinary education."

THE PRACTITIONER.—"It has distinct merit and utility as a clear and readable account of an important subject, treated from many points of view."

Pott 8vo. 1s.

Primer of Hygiene. By ERNEST S. REYNOLDS, M.D. Lond., Member of the Royal College of Physicians of London, Senior Physician to the Ancoats Hospital, Manchester, Victoria University Extension Lecturer on Hygiene, etc., etc. With 50 Illustrations.

THE BRITISH MEDICAL JOURNAL.—"This little work is chiefly for the use of school children, or for persons whose knowledge of the principles of hygiene may be considered extremely limited. . . . . It may be confidently recommended for the purpose it is intended."

THE LANCET.—"It contains information in a style suitable for higher grade school children—lucid, concise, interesting, and useful. . . . . This primer will fulfil the purpose, and the matter it contains will amply satisfy the requirements of the South Kensington examinations."

THE HOSPITAL.—"A great deal of valuable information is conveyed in the space of 150 small pages."

Globe 8vo. 2s. 6d.

Hygiene for Beginners. By ERNEST S. REYNOLDS, M.D. Lond., Member of the Royal College of Physicians of London, Senior Physician to the Ancoats Hospital, Manchester, Victoria University Extension Lecturer on Hygiene, etc., etc.

THE LANCET.—"This is an excellent little book, and one we can recommend to parents, guardians, schoolmasters, and others."

THE HOSPITAL.—"This is a most excellent little book of its kind, and we can commend its use to both students and teachers of elementary hygiene. . . . We trust it may have the appreciation and circulation it undoubtedly deserves."

NATURE.—"We venture to say that every householder, and every mother having the care of children, should be acquainted with as much of the elementary principles of hygiene as is contained in this volume."

Crown 8vo. 8s. 6d.

Diphtheria: Its Natural History and Prevention.

Being the Gilroy Lectures delivered before the Royal College of Physicians of London, 1891, by R. THORNE THORNE, M.B. Lond., F.R.C.P., F.R.S., Assistant Medical Officer to Her Majesty's Local Government Board.

THE BRITISH MEDICAL JOURNAL.—"Dr. Thorne Thorne's scholarly style and lucid analysis of complicated records have never been shown to greater advantage than in the work now before us. . . . Every point in connection with the etiology of diphtheria is impartially reviewed, and the conclusions . . . . are placed before the reader in the clearest light of which the present state of our knowledge permits. Dr. Thorne Thorne's work will at once take its place as the standard authority upon the subject."

THE HOSPITAL.—"It is eminently a book for all fathers and mothers, school teachers and school managers, statesmen, policemen, and sanitary inspectors, and indeed for all sorts and conditions of men and women who have any sort of responsibility for children and young people in either a public or a private capacity."

Crown 8vo. 6s. net.

Vaccination: Its Natural History and Pathology.

Being the Milroy Lectures for 1898, delivered before the Royal College of Physicians of London, by S. MONCKTON COPEMAN, M.A., M.D. Cantab, M.R.C.P. Lond., Medical Inspector to H.M. Local Government Board. With 12 full-page Plates.

Foolscap 8vo. 4s. 6d.

Handbook of Public Health and Demography.

By EDWARD F. WILLOUGHBY, M.B. Lond., Diploma in State Medicine of the London University, and in Public Health of Cambridge University.

THE LANCET.—"We . . . . can honestly recommend it for the scientific accuracy, arrangement, and clearness of treatment with which the various subjects are discussed. There can be but little doubt that public health officers will be armed and well prepared in the principles and practice of hygiene if they are in possession of this handbook."

THE BRITISH MEDICAL JOURNAL.—"The book contains much valuable information."

THE EDUCATIONAL NEWS.—"It is a concise, readable, well stated, and excellently reasoned compendium."

MACMILLAN & CO., LTD.,
ST. MARTIN'S STREET, LONDON, W.C.

2.1.99.

www.ingramcontent.com/pod-product-compliance
Lightning Source LLC
Chambersburg PA
CBHW032002230426
43672CB00010B/2239